Lecture Notes in Electrical Engineering

Volume 317

About this Series

"Lecture Notes in Electrical Engineering (LNEE)" is a book series which reports the latest research and developments in Electrical Engineering, namely:

- Communication, Networks, and Information Theory
- Computer Engineering
- Signal, Image, Speech and Information Processing
- Circuits and Systems
- Bioengineering

LNEE publishes authored monographs and contributed volumes which present cutting edge research information as well as new perspectives on classical fields, while maintaining Springer's high standards of academic excellence. Also considered for publication are lecture materials, proceedings, and other related materials of exceptionally high quality and interest. The subject matter should be original and timely, reporting the latest research and developments in all areas of electrical engineering.

The audience for the books in LNEE consists of advanced level students, researchers, and industry professionals working at the forefront of their fields. Much like Springer's other Lecture Notes series, LNEE will be distributed through Springer's print and electronic publishing channels.

More information about this series at http://www.springer.com/series/7818

Bijnan Bandyopadhyay · Shyam Kamal

Stabilization and Control of Fractional Order Systems: A Sliding Mode Approach

Springer

Bijnan Bandyopadhyay
Shyam Kamal
Interdisciplinary Programme in Systems
and Control Engineering
Indian Institute of Technology Bombay
Mumbai
India

ISSN 1876-1100 ISSN 1876-1119 (electronic)
ISBN 978-3-319-38133-6 ISBN 978-3-319-08621-7 (eBook)
DOI 10.1007/978-3-319-08621-7

Springer Cham Heidelberg New York Dordrecht London

Softcover reprint of the hardcover 1st edition 2015

Printed on acid-free paper

Springer is part of Springer Science+Business Media (www.springer.com)

Saraswati Vandana Mantra is an important hindu mantra that is recited for higher knowledge and wisdom. The goddess Saraswati is the authority on academics and the arts. Everybody from musicians to scientists following Hinduism pray to her for guidance and knowledge.
The Saraswati Vandana Mantra is recited by her devotees every morning for good luck. Everyone has a different version for the vandana which just means holy song. So as a student you ask for physical knowledge as a musician ask for Timing Taals etc.

Mantra

Yaa Kundendu tushaara haara-dhavalaa,
Yaa shubhra-vastra'avritaa
Yaa veena-vara-danda-manditakara,
Yaa shweta padma'asana
Yaa brahma'achyuta shankara prabhritibhir,
Devai-sadaa pujitaa
Saa Maam Paatu Saraswati Bhagavatee
Nihshesha jaadya'apahaa.
Shuklaam Brahmavichaara Saara paramaam
Aadhyaam Jagadvyapinim,

Veena Pustaka Dhaarineem Abhayadaam
Jaadya'andhakaara'apahaam
Haste Sphaatika Maalikam Vidadhateem
Padmasane Sansthitaam
Vande taam Parmeshwareem Bhagavateem
Buddhipradaam Shardam.

English Translation

She, who is as fair as the Kunda flower,
white as the moon, and a garland
of Tushar flowers and who is covered in
white clothes. She, whose hands are
adorned by the excellent veena,
and whose seat is the pure white lotus.
She, who is praised by Brahma,
Vishnu, and Mahesh and prayed
to by the Devas. O Mother Goddess,
remove my mental dullness!

Rigveda (http://en.wikipedia.org/wiki/Saraswati_Vandana_Mantra)

To our Parents, Teachers, Family and Friends, who made us capable enough to write this book

Preface

We will like to start our story by paying our tributes to the book by the name "The Calculus of Friendship [1]" written by "Steven Strogatz."

> Calculus thrives on continuity. At its core is the assumption that things change smoothly, that everything is only infinitesimally different from what it was a moment before. Like a movie, calculus reimagines reality as a series of snapshots, and then recombines them, instant by instant, frame by frame, the succession of imperceptible changes creating an illusion of seamless flow. This way of understanding a change has proven to be powerful beyond words, perhaps the greatest idea that humanity has ever had.
>
> Yet in another way, calculus is fundamentally naive, almost childish in its optimism. Experience teaches us that change can be sudden, discontinuous, and wrenching. Calculus draws its power by refusing to see that. It insists on a world without accidents, where one thing leads logically to another. Give me the initial conditions and the law of motion, and with calculus I can predict the future or better yet, reconstruct the past.

It is quite obvious from the above lines that whenever some infinitesimal change occurs, we have to have calculus as a tool to capture this phenomenon. The thing to be considered is "why we want to capture it?"

And even the answer is quite obvious that we are interested to extrapolate past and future behavior. Therefore, the quality and reliability of the information directly depends on the versatility of the calculus used. Whenever we talk about versatility of the calculus, the scope of Newtonian Calculus becomes limited as its operators are defined for the integer order points only. The other limiting property of this calculus is that the differential operator which is used to capture the phenomenon has only local property. It means that this differential operator is not able to capture the memory-dependent phenomenon.

However, most of the system has memory and so extrapolations of past and future behavior from the present does not only depend on that information at the local moment but also on the memory. Moreover this memory is not always uniform; it is always time- and event-dependent phenomenon.

Therefore, we have to find some suitable nonlocal differential operators which are able to capture both memory-dependent and independent practical phenomenon. The bottom line is we need more generalized calculus of which

Newtonian calculus is just a special case. Coincidentally with Newtonian Calculus, there was a parallel development of another stream of calculus called Fractional Calculus.

The idea of fractional calculus has been known since the development of the regular integer order calculus with the initial works being associated with Leibniz and L'Hospital. In spite of being such an old topic, the developments in this field were rather slow especially if we compare with the integer order "calculus." One of the reasons that might have attributed to this was the absence of any widely accepted geometrical and physical interpretation of the fractional differentials. This limited the concept to just a theory. But lately, there has been a steady growth of interest in this field as it has been realized that the fractional calculus can describe certain phenomenon much better than their integer order counter-parts.

A number of papers by Ross [2–5] and recently by Machado et al. [6, 7] deal with various aspects of the history of fractional calculus. Same is also discussed here very briefly.

When we look back the history of fractional calculus, then it is fond that the first reference probably being associated with a letter exchanged between Leibniz and L'Hospital in 1695 where half-order derivative was mentioned. Further in a short correspondence between Johann Bernoulli and Leibniz in 1695, Leibniz mentioned the derivative of an arbitrary order. In 1730, the subject of fractional calculus was brought to Euler's attention. J. L. Lagrange in 1772 contributed the law of exponents for differential operators, which was further analyzed as the indirect contribution to the fractional calculus. In 1812, P. S. Laplace defined the fractional derivative by means of Cauchy integral and in 1819, S. F. Lacroix mentioned a derivative of arbitrary order in his 700-page long text, followed by J. B. J. Fourier in 1822, who mentioned the derivative of an arbitrary order.

Most important aspect of fractional operator is its application; the earlier usage of the same can be associated with N. H. Abel in 1823 for the solution of Tautochrome problem. J. Liouville made the first major study of fractional calculus in 1832, where he applied his definitions to problems in theory. In 1867, A. K. Grunwald worked on the fractional operations. G. F. B. Riemann developed the theory of fractional integration during his school days and published his paper in 1892. A. V. Letnikov wrote several papers on this topic from 1868 to 1872. Oliver Heaviside published a collection of papers in 1892, where he showed the so-called Heaviside operational calculus is concerned with linear generalized operators. In the period of 1900–1970 the principal contributors to the subject of fractional calculus were, for example, H. H. Hardy, S. Samko, H. Weyl, M. Riesz, S. Blair, etc. From 1970 to the present, the major contributors include J. Spanier, K. B. Oldham, B. Ross, K. Nishimoto, O. Marichev, A. Kilbas, H. M. Srivastava, R. Bagley, K. S. Miller, M. Caputo, I. Podlubny, and many others.

In recent years a number of books [8–39] etc., which are related to fractional calculus and its applications were published in the literature. In these books, almost all the primary theories of fractional order development and its growing applications were discussed. However, if we talk about stability and stabilization of fractional order systems, it is still a growing field and much work is required in this area.

After a survey paper by Utkin in the late 1970s, sliding mode control methodologies emerged as an effective tool to tackle uncertainty and disturbances which are inevitable in most of the practical systems. Sliding mode control is a particular class of variable structure control which was introduced by Emel'yanov and his colleagues. The design paradigms of sliding mode control have now become a mature design technique for the design of robust controller for uncertain system.

In the last two decades, fractional differential equations have been used more frequently to model various physical phenomena. In fact, recent advances of fractional calculus are dominated by modern examples of applications in differential and integral equations, physics, signal processing, fluid mechanics, viscoelasticity, mathematical biology, electro chemistry and many others. We can refer to [40] for the recent history of fractional calculus. Fractional calculus is steadily becoming an exciting new mathematical method of solution of diverse problems in mathematics, science, and engineering.

As a result of the growing applications, the study of stability of fractional differential equations has attracted much attention [41–44]. Furthermore, in recent years, an increasing attention has been given to fractional order controllers, and many progresses are reflected in [45–47] and references cited therein. Some of these applications include optimal control [48, 49], CRONE controller [50], fractional PID controller [51, 52], and lead-lag compensator [53].

The term "non-integer calculus" is sometimes used instead of "fractional calculus" in literature. However, the fractional calculus covers integer orders as well as fractional, irrational, complex, and generalized functional orders also. For this reason, a third name was even coined: "generalized calculus." All these names are used interchangeably in this monograph.

Sliding mode controller [54] and VSC controller have been already implemented to improve the performance and robustness in closed loop control systems. Sliding mode control (SMC) has many attractive features such as invariance to matched uncertainties, order reduction, simplicity in design, robustness against perturbations, and some others [55], [54] and [56]. The characteristic feature of continuous time SMC system is that sliding mode occurs on a prescribed manifold (sliding surface), where switching control is employed to maintain the states on the surface [57].

Sliding mode control has two phases (a) reaching phase in which the system states are driven from any initial state to reach the switching manifolds (the anticipated sliding modes) in finite time and (b) sliding phase in which the system is induced into the sliding motion on the switching manifolds, i.e., the switching manifolds become attractors. The robustness and order reduction property of sliding mode control comes into picture only after the occurrence of sliding mode. During the reaching phase, however, there is no guarantee of the above properties [54]. Integral sliding mode [58, 59] though is a variant. It eliminates the reaching phase by enforcing the sliding mode throughout the entire system response.

In the field of fractional order systems, the application of SMC was reported only recently. In these works, the philosophy of integer order SMC was extended to fractional order systems [60, 61]. The SMC design for fractional order systems with input and state delays was also proposed [62]. The application of SMC to fractional order system has been reported in [63–67]. In the above cited references, only asymptotic stabilization of states could be achieved. The most recent attempt at combining SMC with fractional order system was done in [68] in which super-twisting control was used. But there also, only asymptotic stabilization could be achieved.

All the developments mentioned above are for the continuous fractional order systems. However, a large class of continuous time systems is controlled by computers and microcontrollers. In these cases, information about the system is available only at specific time instants and control inputs can only be applied at these instants. Due to this, discrete modeling of systems and controllers is more relevant than their continuous counterpart. Not much work is reported in the area of discrete-time fractional order modeling, stabilization, and control. A few results are available though, but they are applicable for restricted class of discrete fractional systems. These are discussed in the recent book [69], monograph [31], and references within. State space representation of the linear discrete-time fractional order system has been introduced in [70]. Using this representation, observers, Kalman filters and feedback controllers are designed for the discrete fractional order systems [71–73]. Some stability conditions have been derived in [74]. Observability and controllability properties of discrete fractional order system are discussed in [72].

To implement the continuous time variable structure control algorithm sampled-data environment is required, but in many practical places one cannot achieve performance one would expect on the basis of the continuous time theory. Hence, in such a situation, the option of discrete-time sliding mode comes into picture. Also relatively low switching frequency is required than the continuous time sliding mode control (theoretically infinite frequency is required to maintain trajectories on the surface), so, discrete control algorithm is more practical to implement. In case of discrete sliding mode (DSM) design, the control input is computed only at certain sampling instants and the control effort is held constant over the entire sampling period.

The two schools of thought on discrete-time sliding mode control are (1) by using a switching-based control law and (2) without using any switching function in the control law. In [75], Gao et al. has used the switching-based reaching law, which shows that the motion of a discrete SMC system can undergo only quasi-sliding mode. It is also known that, chattering occurs in the system due to the application of a switching control [76]. It has been shown in [75], that due to the use of the switching function, the system states would reach the vicinity of the origin, but cannot get arbitrarily close to the origin. The other one is nonswitching type. It has been shown in [77] that, sliding mode may be achieved in discrete-time systems without the usage of a switching function. This is due to the fact that, discrete-time control is inherently discontinuous in nature and thus may not require an explicit discontinuity in the control law to bring out sliding mode control. Such a sliding mode control without chattering was proposed in [78]. Here, the control law uses apriori known function to lead the system states onto the sliding surface. The discrete-time sliding mode control using an equivalent control is proposed by Utkin [76]. The control does not need the switching function and it brings the state trajectory to the sliding surface in one sampling time. With limited control, trajectory reaches the sliding surface in finite sampling steps. To the best of author's knowledge, sliding mode control of fractional discrete-time system is not achieved before the authors' work.

One of the unexplored area in fractional order is soft variable structure control [79–81]. It is a class of variable structure control in which, controller parameters or structures are continuously varying or switching. Due to continuous switching between different controllers, it has many advantages such as achieving high regulation rates, shortening the settling time, and hardly any system chattering. We can refer to [79] and references within that for the detailed history, definition and unsolved problems of integer order soft VSC. In soft variable structure actually Lyapunov function depends on the behavior of states and switchs both. Therefore, high regulation rate and short settling time comes into picture. Due to the above properties the term "fast asymptotic stability" is used, which is common in soft variable structure [79]. To the best of author's knowledge, soft VSC of fractional order system is not reported before the authors' work.

There are several types of stability concepts proposed in control theory depending on how and when the system is stabilized. One of the most classic concepts regarding stability is obtained by introducing a weighted norm, or more generally by defining a Lyapunov function, which is decreasing at each instant and is strongly related to the norm of the system states. However, the appropriate Lyapunov functions are not always easy to find, because of unavailability of some concrete approach.

The above limitation can be overcome if the convergence problem is interpreted as a property of all solution converge toward one another. The concept of contraction leads to the introduction of suitable Riemann metrics or more generally Finsler metrics. Recently, inspired from fluid mechanics and differential geometry, Lohmiller and Slotine proposed a new method of stability analysis known as contraction theory [82, 83].

However, classical contraction theory which is proposed in literature [82, 83], able to analyze the convergence behavior of nonlinear systems in state space only when the system is continuously differentiable. But, lot of systems existing in real-time applications where systems are modeled using nondifferentiable equations. In the present work, we revisited the contraction theory by replacing the first-order variation of system state by the fractional order variation. Classical contraction theory is able to analyze the convergence behavior of nonlinear systems in state space only when the system is continuously differentiable. The main advantage of the proposed approach is that it also works for analyzing the stability of nondifferentiable systems. Also, this approach is useful for analyzing stability of fractional order systems and designing the fractional order controller which is better than the integer order controller for dynamical systems described by integer order differential equation.

Motivation

In numerous applications, it is required that the control objective is achieved in finite time. In the case of finite-time stabilization, as the name suggests the system states reach the system equilibrium in a finite time [84]. It is required for achieving faster convergence, better robustness as well as disturbance rejection properties. It also gives the optimality in settling time of the controlled system [85, 86]. In primary development, finite-time convergence property was considered to be only for nonsmooth or non-Lipschitz continuous autonomous system [85].

Recently, finite-time stability has been further extended to nonautonomous systems [87], switched systems [88], time-delay systems [89, 90], impulsive dynamical systems [91], and stochastic nonlinear systems [92]. In the case of fractional order systems, [93] discusses point to point control of fractional order systems in which any initial state is transferred to any final state in a specified finite amount of time by using an open-loop control. However, being an open-loop control it is limited to a disturbance free environment. Our current work proposes a novel and simple solution to the addressed problem.

Further, the higher order sliding mode control for integer order case with respect to output is equivalent to the finite-time stabilization to zero of integrator chain with nonlinear uncertainties. Main motivation behind the above work is to extend this concept for fractional order systems. Therefore, we propose a novel methodology for robust finite-time stabilization of fractional order systems using integral sliding mode algorithm. An improved strategy with more general kind of uncertainty is also proposed. Since, sliding mode control is used, the proposed controller makes the system robust to matched uncertainties.

As discussed in the introduction part, a large class of continuous time systems is controlled by computers and microcontrollers. In these cases, information about the system is available only at specific time instants and control inputs can only be applied at these instants. Motivating from this fact, we define discrete fractional order sliding mode control to achieve robustness with respect to matched uncertainties which is not possible using simple state feedback.

Short settling and high regulation rate is one of the prime issues for the fractional order system. Inspired from this fact, we developed soft variable structure control for fractional order systems. Due to the alleviation of chattering phenomenon soft VSC that is proposed for fractional order LTI systems are quite useful in future for the practical systems.

Apart from the above this monograph, revisited the contraction theory by replacing the first-order variation of system state by the fractional order variation

[94, 95]. The main advantage of the proposed approach is that it also works for analyzing the stability of nondifferentiable systems. Also, this approach is useful for analyzing the stability of fractional order systems and designing the fractional order controller which is better than the integer order controller for dynamical systems described by integer order differential equation [96].

The Monograph

The monograph is based on the authors' work on, stabilization and control design for continuous and discrete fractional order systems. Initial two chapters and some parts of third chapter of this monograph are written in tutorial fashion to cover all the basic concepts of fractional order system and brief overview of sliding mode control of fractional order system. In Chap. 2, lot of stress is given to what is the need for sliding mode control of fractional order system? How to define solution of fractional order differential equation with discontinuous right hand side? What are the different approaches to synthesize reaching law for fractional order system? How to design sliding surface? The rest part contains the contributory chapters of the monograph may be summarized in the following manner:

- In Chap. 3, a novel method for finite-time stabilization of a chain of uncertain fractional order integrator is proposed for the first time. This is accomplished by first designing a controller which is capable of stabilizing the disturbance free states of the system in finite time. Then a suitable sliding surface is designed such that when the system slides on it, the designed controller is transferred to act on the disturbance free states. After that, by switched sliding surface mechanism the remaining state which is affected by the disturbance is also stabilized.
- Higher order sliding mode control of fractional order system is formulated in terms of output and their chain of fractional derivatives in Chap. 4. This is achieved by first designing a nominal controller which stabilizes the system in finite time. An integral sliding mode like surface and a switching controller is proposed such that when the system is on the surface, the equivalent value of the integral sliding mode control is the negative of the disturbance and hence the disturbance is cancelled. An improved strategy with more general kind of uncertainty is also proposed.
- In Chap. 5, soft variable structure control of fractional order systems is proposed for achieving high regulation rates and short settling times. Using the appropriate fractional order Lyapunov approach, a bilinear switching-based controller and a controller based on continuous fractional switch are designed to achieve high control performance.
- The Chap. 6, deals with the problem of cooperative control of networked fractional order multiagent systems over a directed interaction graph. For achieving the specified goal, a new fractional order continuous control law is

designed based on sliding mode theory. finite-time reachability to the sliding surface is proved using fractional order extension of Lyapunov stability criterion.
- In Chap. 7, a methodology for stabilization of fractional discrete-time system based on discrete sliding mode approach is attempted. The proposed controller is robust in the presence of matched uncertainties. Stability during sliding is analyzed in terms of gamma function-based radius using the property introduced by Hilfer.
- The Chap. 8, presents one of the open problems, the robust controller design for discrete fractional order systems. For achieving the specified goal a disturbance observer is constructed based on full state information. Then, an extra compensator based on the theory of disturbance observer is added in the classical state feedback. It is also shown that after applying the proposed controller the closed loop system would become robust against the disturbance as the estimation error becomes small.
- There are several types of stability concepts proposed in control theory depending on how and when the system is stabilized. One of the most classic concepts regarding stability is obtained by introducing a weighted norm, or more generally by defining a Lyapunov function, which is decreasing at each instant and is strongly related to the norm of the system states. The revisiting concept of contraction leads to the introduction of suitable Riemann metrics or more generally Finsler metrics. Recently, inspired from fluid mechanics and differential geometry, Lohmiller and Slotine proposed a new method of stability analysis known as contraction theory [85, 84].

This theory is based on the concept that the stability can be analyzed differentially by analyzing if the nearby trajectories converge to one another, rather than through finding some implicit motion integral as in Lyapunov theory, or through some global state transformation as in feedback linearization. Making use of the concepts of contraction theory in Chap. 9, we have attempted to design a globally exponentially stable controller for fractional order systems.

Also in this chapter, the contraction theory is revisited by replacing the first-order variation of system state by the fractional order variation. Classical contraction theory is able to analyze the convergence behavior of nonlinear systems in state space only when the system is continuously differentiable. The main advantage of the proposed approach is that it also works for analyzing the stability of nondifferentiable systems. Also, this approach is useful for analyzing stability of fractional order systems and designing the fractional order controller which is perhaps better than the integer order controller for dynamical systems described by integer order differential equation.

The purpose of this monograph is to give a different dimension to robust control of fractional order systems. While completing this monograph, we realized that it is not only of authors' but of many! We acknowledge all those many!!

God has given us the strength, patience, and courage to sail across in spite of many constraints. We acknowledge the great source of power. SYSCON has a wonderful research culture. The useful discussions among the researchers, helping attitude of the research scholars are memorable.

The authors would like to express their deep sense of gratitude to their parents and teachers who have made them capable enough to write this book. The authors wish to thank many individuals who had helped them directly or indirectly in completing this monograph. In particular, the authors wish to thanks their friends and colleagues in Systems and Control group, (SYSCON), IIT Bombay. Finally authors wish to acknowledge the support, patience, and love of their family and friends during the preparation of this monograph.

Bombay, India, May 2014

Bijnan Bandyopadhyay
Shyam Kamal

References

1. Strogatz, S.: The calculus of friendship. Thought & Action 39 (2009)
2. Ross, B.: Fractional calculus and its applications. In: Proceedings of the international conference held at the University of New Haven, June, 1974, vol. 457. Springer, Heidelberg (1975)
3. Ross, B.: The development of fractional calculus 1695–1900. Historia Mathematica **4**(1), 75–89 (1977)
4. Ross, B.: Fractional calculus. Math. Mag. **50**(3), 115–122 (1977)
5. Ross, B.: Origins of fractional calculus and some applications. Int. J. Math. Statist. Sci **1**(1), 21–34 (1992)
6. Machado, J.T., Kiryakova, V., Mainardi, F.: A poster about the recent history of fractional calculus. Frac. Calc. Appl. Anal **13**(3), 329–334 (2010)
7. Tenreiro Machado, J., Kiryakova, V., Mainardi, F.: A poster about the old history of fractional calculus. Fractional Calc. Appl. Anal. **13**(4), 447–454 (2010)
8. Sabatier, J., Agrawal, O.P., Machado, J.T.: Advances in Fractional Calculus: Theoretical Developments and Applications in Physics and Engineering. Springer, New York (2007)
9. Miller, K.S., Ross, B.: An Introduction to the Fractional Calculus and Fractional Differential Equations. John Wiley and Sons, Inc., New York (1993)
10. van Roermund, A.H., Steyaert, M., Huijsing, J.: Analog Circuit Design: Fractional-N Synthesizers, Design for Robustness, Line and Bus Drivers. Springer, Heidelberg (2003)
11. Blei, R.C., et al.: Analysis in Integer and Fractional Dimensions, vol. 71. Cambridge University Press, Cambridge (2001)
12. Hilfer, R.: Applications of Fractional Calculus in Physics. Word Scientific, Singapore (2000)
13. Mainardi, F.: Fractional Calculus and Waves in Linear Viscoelasticity: An Introduction to Mathematical Models. World Scientific, Singapore (2010)
14. Ortigueira, M.D.: Fractional Calculus for Scientists and Engineers, vol. 84. Springer, Heidelberg (2011)
15. Anastassiou, G.A.: Fractional Differentiation Inequalities. Springer, New York (2009)
16. Joseph, K., et al.: Fractional Dynamics: Recent Advances. World Scientific, Singapore (2012)
17. Baleanu, D., Machado, J.A.T., Luo, A.C.: Fractional Dynamics and Control. Springer, New York (2012)
18. Scheinerman, E.R., Ullman, D.H.: Fractional Graph Theory. Dover Publications, New York (2011)

19. Samko, S.G., Kilbas, A.A., Marichev, O.O.I.: Fractional Integrals and Derivatives. Gordon and Breach Science Publishers, Yverdon (1993)
20. Luo, Y., Chen, Y.: Fractional Order Motion Controls. Wiley, New York (2012)
21. Caponetto, R.: Fractional Order Systems: Modeling and Control Applications, vol. 72. World Scientific, Singapore (2010)
22. Sheng, H., Chen, Y., Qiu, T.: Fractional Processes and Fractional-Order Signal Processing. Springer, London (2012)
23. Das, S.: Functional Fractional Calculus. Springer, Berlin (2011)
24. Das, S.: Functional Fractional Calculus for System Identification and Controls. Springer, Berlin (2008)
25. Kirjakova, V.: Generalized Fractional Calculus and Applications, vol. 301. CRC Press, Boca Raton (1994)
26. Zaslavsky, G.M.: Hamiltonian Chaos and Fractional Dynamics, vol. 1. Oxford University Press, Oxford (2005)
27. Nourdin, I.: Selected Aspects of Fractional Brownian Motion. Springer, Milano (2012)
28. Meerschaert, M.M., Sikorskii, A.: Stochastic Models for Fractional Calculus, vol. 43. Walter de Gruyter, Berlin (2011)
29. Abbas, S., Benchohra, M.: Topics in Fractional Differential Equations, vol. 27. Springer, New York (2012)
30. Valerio, D., Da Costa, J.S.: An Introduction to Fractional Control, vol. 91. Institution of Engineering and Technology (IET), England (2012)
31. Kaczorek, T.: Selected Problems of Fractional Systems Theory, vol. 411. Springer, Berlin (2011)
32. Diethelm, K.: The Analysis of Fractional Differential Equations: An Application-Oriented Exposition Using Differential Operators of Caputo Type, vol. 2004. Springer, Heidelberg (2010)
33. Annaby, M.H., Mansour, Z.S.: Q-Fractional Calculus and Equations. Springer, Berlin (2012)
34. Cohen, S., Istas, J.: Fractional Fields and Applications. Springer, New York (2013)
35. Magin, R.L.: Fractional Calculus in Bioengineering, part 1. Critic. Reviews in Biomedical Engineering, vol. 32(1). (2004)
36. Das, S., Pan, I.: Fractional Order Signal Processing: Introductory Concepts and Applications. Springer, Heidelberg (2012)
37. Anastassiou, G.A.: Advances on Fractional Inequalities. Springer, New York (2011)
38. Pan, I., Das, S.: Intelligent Fractional Order Systems and Control: An Introduction. Springer (2013)
39. De Almeida, M.F., Ferreira, L.C.: Self-similarity, symmetries and asymptotic behavior in morrey spaces for a fractional wave equation. Diff. Integr. Eqn. **25**(9/10), 957–976 (2012)
40. Machado, J.T., Kiryakova, V., Mainardi, F.: Recent history of fractional calculus. Commun. Nonlinear Sci. Numer. Simul. **16**(3), 1140–1153 (2011)
41. Agrawal, O.P., Muslih, S.I., Baleanu, D.: Generalized variational calculus in terms of multiparameters fractional derivatives. Commun. Nonlinear Sci. Numer. Simul. **16**(12), 4756–4767 (2011)
42. Agrawal, O.P.: Fractional variational calculus in terms of riesz fractional derivatives. J. Phys. A: Math. Theor. **40**(24), 6287 (2007)
43. Herzallah, M.A., Baleanu, D.: Fractional euler–lagrange equations revisited. Nonlinear Dyn. **69**(3), 977–982 (2012)
44. Li, C., Zhang, F.: A survey on the stability of fractional differential equations. Eur. Phys. J. Spec. Top. **193**(1), 27–47 (2011)
45. Diethelm, K., Scalas, E.: Fractional Calculus: Models and Numerical Methods, vol. 3. World Scientific Publishing Company, Hackensack (2012)
46. Ortigueira, M.D.: An introduction to the fractional continuous-time linear systems: the 21st century systems. IEEE Circuits Syst. Mag. **8**(3), 19–26 (2008)

47. Xue, D., Chen, Y.: A comparative introduction of four fractional order controllers. In: Intelligent Control and Automation, 2002. Proceedings of the 4th World Congress on, vol. 4, Shanghai, P.R.China, 10–14 June, pp. 3228–3235. IEEE (2002)
48. Agrawal, O.P.: A general formulation and solution scheme for fractional optimal control problems. Nonlinear Dyn. **38**(1–4), 323–337 (2004)
49. Agrawal, O.P., Baleanu, D.: A Hamiltonian formulation and a direct numerical scheme for fractional optimal control problems. J. Vib. Control **13**(9–10), 1269–1281 (2007)
50. Oustaloup, A., Mathieu, B., Lanusse, P.: The crone control of resonant plants: application to a flexible transmission. European J. Control **1**(2), 113–121 (1995)
51. Bouafoura, M.K., Braiek, N.B.: Piλdμ controller design for integer and fractional plants using piecewise orthogonal functions. Commun. Nonlinear Sci. Numer. Simul. **15**(5), 1267–1278 (2010)
52. Podlubny, I.: Fractional-order systems and pi/sup/spl lambda//d/sup/spl mu//-controllers. Automatic Control, IEEE Trans. **44**(1), 208–214 (1999)
53. Raynaud, H.F., Zergaınoh, A.: State-space representation for fractional order controllers. Automatica **36**(7), 1017–1021 (2000)
54. Utkin, V., Guldner, J., Shi, J.: Sliding Mode Control in Electro-Mechanical Systems. CRC Press, Boca Raton (2009)
55. Edwards, C., Spurgeon, S.K.: Sliding Mode Control: Theory and Applications, vol. 7. CRC Press, Boca Raton (1998)
56. Utkin, V.I.: Sliding Modes in Control and Optimization, vol. 116. Springer-Verlag, Berlin (1992)
57. Fridman, L., Levant, A.: Higher order sliding modes. Sliding Mode Control Eng. **11**, 53–102 (2002)
58. Cao, W.J., Xu, J.X.: Nonlinear integral-type sliding surface for both matched and unmatched uncertain systems. IEEE Trans. Autom. Control **49**(8), 1355–1360 (2004)
59. Castaños, F., Fridman, L.: Analysis and design of integral sliding manifolds for systems with unmatched perturbations. IEEE Trans. Autom. Control **51**(5), 853–858 (2006)
60. Efe, M.O.: Fractional order sliding mode control with reaching law approach. Turk. J. Electr. Eng. Comput. Sci **18**(5), 731–747 (2010)
61. Efe, M.Ö.: Fractional order sliding mode controller design for fractional order dynamic systems. In: New Trends in Nanotechnology and Fractional Calculus Applications, pp. 463–470. Springer, New york (2010)
62. Si-Ammour, A., Djennoune, S., Bettayeb, M.: A sliding mode control for linear fractional systems with input and state delays. Commun. Nonlinear Sci. Numer. Simul. **14**(5), 2310–2318 (2009)
63. Bai, J., Yu, Y.: Sliding mode control of fractional-order hyperchaotic systems. In: Chaos-Fractals Theories and Applications (IWCFTA), 2010 International Workshop on Kunming, Yunnan, China, 29–31 October 2010, pp. 111–115. IEEE (2010)
64. Chang, Y.H., et al.: Fractional-order integral slidingmode flux observer for sensorless vector-controlled induction motors. In: American Control Conference (ACC), 2011, pp. 190–195. IEEE (2011)
65. Hosseinnia, S., et al.: Sliding mode synchronization of an uncertain fractional order chaotic system. Comput. Math. Appl. **59**(5), 1637–1643 (2010)
66. HosseinNia, S.H., et al.: Augmented system approach for fractional order smc of a dc–dc buck converter. In: Proceedings of IFAC Workshop on Fractional Differentiation and its Applications, pp. 18–20 (2010)
67. Ramírez, H.S., Battle, V.F.: A generalized pi sliding mode and pwm control of switched fractional systems. In: Modern Sliding Mode Control Theory, pp. 201–221. Springer (2008)
68. Pisano, A., Rapaíc, M., Usai, E.: Second-order sliding mode approaches to control and estimation for fractional order dynamics. In: Sliding Modes after the First Decade of the 21st Century, pp. 169–197. Springer, Berlin Heidelberg (2012)

69. Baleanu, D., Güvencù, Z.B., Machado, J.T.: New trends in nanotechnology and fractional calculus applications. Springer, New york (2010)
70. Raynaud, H.F., Zergaınoh, A.: State-space representation for fractional order controllers. Automatica **36**(7), 1017–1021 (2000)
71. Dzielinski, A., Sierociuk, D.: Adaptive feedback control of fractional order discrete state space systems. In: Computational Intelligence for Modelling, Control and Automation, 2005 and International Conference on Intelligent Agents, Web Technologies and Internet Commerce, International Conference on, vol. 1, pp. 804–809. IEEE (2005)
72. Guermah, S., Djennoune, S., Bettayeb, M.: Controllability and observability of linear discrete-time fractional-order systems. Int. J. Appl. Math. Comput. Sci. **18**(2), 213–222 (2008)
73. Sierociuk, D., Dzielínski, A.: Fractional kalman filter algorithm for the states, parameters and order of fractional system estimation. Int. J. Appl. Math.Comput. Sci. **16**, 129–140 (2006)
74. Dzielínski, A., Sierociuk, D.: Stability of discrete fractional order state-space systems. J. Vib. Control **14**(9–10), 1543–1556 (2008)
75. Gao, W., Wang, Y., Homaifa, A.: Discrete-time variable structure control systems. IEEE Trans. Industr. Electron. **42**(2), 117–122 (1995)
76. Utkin, V.: Variable structure systems with sliding modes. IEEE Trans. Autom. Control **22**(2), 212–222 (1977)
77. Bartolini, G., Ferrara, A., Utkin, V.I.: Adaptive sliding mode control in discrete-time systems. Automatica **31**(5), 769–773 (1995)
78. Bartoszewicz, A.: Discrete-time quasi-sliding-mode control strategies. IEEE Trans. Industr. Electron. **45**(4), 633–637 (1998)
79. Adamy, J., Flemming, A.: Soft variable-structure controls: a survey. Automatica **40**(11), 1821–1844 (2004)
80. Su, J.P., Lee, T.E., Yu, K.W.: A combined hard and soft variable-structure control scheme for a class of nonlinear systems. IEEE Trans. Industr. Electron. **56**(9), 3305–3313 (2009)
81. Wang, Y., Xiong, Z., Ding, H.: Fast response and robust controller based on continuous eigenvalue configurations and time delay control. Robot. Computr. Integr. Manuf. **23**(1), 152–157 (2007)
82. Aylward, E.M., Parrilo, P.A., Slotine, J.J.E.: Stability and robustness analysis of nonlinear systems via contraction metrics and sos programming. Automatica **44**(8), 2163–2170 (2008)
83. Lohmiller, W., Slotine, J.J.E.: On contraction analysis for non-linear systems. Automatica **34**(6), 683–696 (1998)
84. Ryan, E.: Singular optimal controls for second-order saturating systems. Int. J. Control **30**(4), 549–564 (1979)
85. Bhat, S.P., Bernstein, D.S.: Finite-time stability of continuous autonomous systems. SIAM J. Control Optim. **38**(3), 751–766 (2000)
86. Haimo, V.: Finite time controllers. SIAM J. Control and Optim. **24**(4), 760–770 (1986)
87. Moulay, E., Perruquetti, W.: Finite time stability conditions for non-autonomous continuous systems. Int. J. Control **81**(5), 797–803 (2008)
88. Orlov, Y.: Finite time stability and robust control synthesis of uncertain switched systems. SIAM J. Control Optim. **43**(4), 1253–1271 (2004)
89. Lazarević, M.: Finite time stability analysis of pd α fractional control of robotic time-delay systems. Mech. Res. Commun. **33**(2), 269–279 (2006)
90. Moulay, E., et al.: Finite-time stability and stabilization of time-delay systems. Syst. Control Lett. **57**(7), 561–566 (2008)
91. Nersesov, S.G., Haddad, W.M.: Finite-time stabilization of nonlinear impulsive dynamical systems. Nonlinear Analy. Hybrid Syst. **2**(3), 832–845 (2008)
92. Chen, W., Jiao, L.: Finite-time stability theorem of stochastic nonlinear systems. Automatica **46**(12), 2105–2108 (2010)

93. Andrzej, D., Wiktor, M.: Point to point control of fractional differential linear control systems. Advances in Differential Equations (2011). Available: www.advancesindifference equations.com/content/2011/1/13
94. Tarasov, V.E.: Fractional variations for dynamical systems: Hamilton and Lagrange approaches. J. Phys. A: Math. Gen. **39**(26), 8409 (2006)
95. Tarasov, V.E.: Fractional derivative as fractional power of derivative. Int. J. Math. **18**(03), 281–299 (2007)
96. Chen, Y.Q., Petras, I., Xue, D.: Fractional order control-a tutorial. In: American Control Conference, 2009. ACC'09, pp. 1397–1411 (2009)

Contents

Acronyms

Abbreviations

CRONE	Commande Robuste d'Ordre Non Entrier
DLN	Domino Ladder Network
DOB	Disturbance Observer
DSM	Discrete-time Sliding Mode
DSMC	Discrete-time Sliding Mode Control
FOS	Fractional Order Systems
IOS	Integer Order Systems
ISM	Integral Sliding Mode
ISMC	Integral Sliding Mode Control
LMI	Linear Matrix Inequality
LTI	Linear Time Invariant
LTV	Linear Time Varying
MIMO	Multi Input Multi Output
ODE	Ordinary Differential Equation
PID	Proportional Integral Derivative
PSW	Piecewise Smooth Systems
QSM	Quasi Sliding Mode
QSMB	Quasi Sliding Mode Band
RL	Riemann–Liouville
VSC	Variable Structure Control

Symbols

$\mathbb{N}$	Set of positive integers
$\mathbb{R}(z) > 0$	Right half of the complex plane
${}_0^{RL}D_t^{\alpha}f(t)$	Riemann–Liouville fractional order derivative
${}_0^{C}D_t^{\alpha}f(t)$	Caputo fractional order derivative

$D^{\alpha}f(t)$	Used for both Caputo/RL fractional order derivative
I	Identity operator
B	Input matrix of Continuous time LTI system
C	Output matrix
x	State vector
u	Control input
z	State vector in regular form
c^T	Sliding surface parameters
ρ, d	System uncertainty/disturbances
c_1, c_2	Sliding surface parameters
$A_{11}, A_{12}, A_{21}, A_{22}$	System matrices in regular form
τ, h	Sampling time
η	Positive constant
$\mathbb{R}$	The field of real numbers
$\mathbb{R}^n$	The real vector space of dimension-*n*
Φ	System matrix of discrete-time LTI system
Γ	Input matrix of discrete-time LTI system
T	Regular form transformation matrix
d	Disturbances/uncertainty
S	Switching function
$H(t)$	Heaviside unit step function
$\mathbb{C}$	Complex number
Re	Real part of complex number
H_1, H_2	Positive definite Hermitian matrices
$\omega_0(t)$	Gain crossover frequency
φ_m	Phase margin
$u_0(t)$	Unit step input
$y_i(t)$	Unit-impulse response
$y_s(t)$	Unit-step response
$\dot{x} \in F(x)$	Filippov differential inclusion
$\overline{co}M$	Convex closure of *M*
$O_{\delta}(x)$	The δ-vicinity of x
f_{ε}	Locally Lipschitzen map
$(.)^*$	Complex conjugate transpose
D	Part of complex left-half plane
$\mathbb{N}_+$	Positive integer
V	Lyapunov function (positive definite)
e	Error between actual and desired trajectory
z_d	Desired trajectory
A_{δ}	System matrix represented with delta operator
B_{δ}	Input matrix represented with delta operator
h	An integer accounts for delay in input
E	Disturbance matrix
$\hat{d}$	Predicted disturbance

$E_{\alpha,\alpha}(.)$	Mittag-Leffler function
$\Gamma(.)$	Gamma function
n	Number of states of a fractional order LTI system model
m	Number of input of a fractional order LTI system model
$\mathrm{eig}(A)$	Eigenvalues of the matrix A
$\mathcal{G} = \{\mathcal{V}, \mathcal{E}\}$	Communication graph between the agents
$\mathcal{V}$	Set of nodes where each node represent an agent
$\mathcal{E}$	Set of edges
$\{\mathcal{V}^s, \mathcal{E}^s\}$	A subgraph of $\{\mathcal{V}, \mathcal{E}\}$
$\Delta^\alpha x(k)$	Grunwald–Letnikov difference of the function $x(k)$
$\begin{pmatrix} \alpha \\ j \end{pmatrix}$	Binomial term
$r(L, \alpha)$	Stability radius of the discrete fractional order system
ε_m	Quasi sliding mode band
$\tilde{x}$	Infinite column sequence
$\tilde{S}$	Backward shift
$\tilde{W}_r^p$	Operator reachability Gramian
$G_{\Xi k}$	Discrete fractional transition matrix
$R(f, f')$	Determinant of the $(2n-1) \times (2n-1)$ Sylvester matrix
$P(\lambda)$	Polynomial equation
$D(P)$	Discriminant of $P(\lambda)$
δx	Virtual displacement (infinitesimal displacement at fixed time)
$M(x(t))$	Symmetric, uniformly positive definite and continuously differentiable metric
$\delta^\alpha x$	Fractional infinitesimal displacement at fixed time

Chapter 1
Essence of Fractional Order Calculus, Physical Interpretation and Applications

1.1 Introduction

In the understanding and development of large class of systems it is now a well realized and accepted fact that the researchers have taken their initiation from nature. Natural things can be well understood in two possible ways viz. quantitative and qualitative. Mathematics plays a central role in this direction. It is the science of patterns and relationships. As a theoretical discipline, mathematics gives a lot of relationships among abstractions without concerning the existence of its counterparts in real world. The abstractions can be anything from strings of numbers to geometric figures to sets of algebraic, difference and differential equations.

When we go back to understand the quantitative and qualitative behavior of nature, it seems that evolution is from integer to fraction. Quantitative behavior can be well explained using number theory, which started from integer and reached to fractional due to division operation and finally converged to real numbers.

Geometry plays an important role to understand the qualitative behavior of the systems, wherein the objects are characterized like point, line, surface and volume, whose dimensions are expressed using the integer number. However, recently it was found that the complete understanding of nature is not possible without the knowledge fractal geometry. In 1983 Mandelbrot pointed out that fractal theory which only provide the true geometry of nature [1].

Calculus is a branch of mathematics describing how things change. It provides a framework for modeling systems undergoing change, and a way to deduce the predictions of such models. Calculus is also developed from integer and extended to fractional or non-integer orders to evolve more generalized calculus. All these resulted in pointing a fact that integer order calculus is a subset of fractional calculus. In this Chapter we are going to discuss a brief overview of the fractional order calculus, physical interpretation and applications.

The brief outline of this chapter is as follows. Section 1.2 describes the formulation of non-integer integral and derivative operators. Section 1.3 discusses the need of fractional calculus in modeling and control of dynamical systems. Some important

B. Bandyopadhyay and S. Kamal, *Stabilization and Control of Fractional Order Systems: A Sliding Mode Approach*, Lecture Notes in Electrical Engineering 317,
DOI 10.1007/978-3-319-08621-7_1

Laplace transform of fractional order operators are analyzed in Sect. 1.4. A deep discussion on the equivalence between fractional order system (FOS) and integer order system (IOS) is presented in Sect. 1.5. Physical realization issues of the fractional derivatives and integral are discussed in Sect. 1.6. Different possible ways of interpretation of fractional integral and derivative are surveyed in Sect. 1.7. Section 1.8 discussed the first application of fractional calculus or operators for physical problem. In the sequences of various applications, Viscoelastic material modeling is discussed in Sect. 1.9, application of fractional calculus in Econophysics is presented in Sect. 1.10, age of the earth using heat equation and fractional order calculus is discussed in Sect. 1.11, dynamical system and fractional order modeling is presented in Sect. 1.12 and how to analyzed the motion of an immersed plate is deeply discussed in the Sect. 1.13. Modeling of dynamical system based on input-output data is one of the most appropriate ways to capture the dynamics of the system. Therefore in Sect. 1.14 data fitting and fractional order modeling is presented. Some more discussion on properties of fractional order operators is discussed in Sect. 1.15. Some more applications of fractional calculus in control is surveyed in Sect. 1.16 followed by the concluding Sect. 1.17.

1.2 Formulation of Non-Integer Integral and Derivative Operators

There are many different methodologies to generalize the integer order integral and derivative operators to non-integer [2, 3]. Possibly the one of the easiest accesses to the idea of the non-integer differential and integral operators studied in the field of fractional calculus is given by Cauchy in this well known representation of an n-fold integral as a convolution integral. This representation is given use the following *Cauchy's formula* [4]

Theorem 1.1 *The n-fold integral of the function $f(t)$ is given as*

$$ {}_0I_t^n f(t) = \int_0^t \frac{(t-\tau)^{n-1}}{(n-1)!} f(\tau) d\tau, \qquad n \in \mathbb{N}, \tag{1.1}$$

where $\mathbb{N}$ represents the set of positive integers.

Proof Consider an antiderivative of the function $f(t)$, represented as ${}_0I_t^1 f(t)$, then one can write

$$ {}_0I_t^1 f(t) = \int_0^t f(x) dx. \tag{1.2}$$

Integrating it one more time giving

$$ {}_0I_t^2 f(t) = \int_0^t \int_0^\tau f(\tau)d\tau dx. \tag{1.3}$$

Equation (1.3) can be considered as a double integral. Using Dirichlet's equality the sequence of integration can be re-expressed by changing the limits. We obtain

$$ {}_0I_t^2 f(t) = \int_0^t \int_\tau^t f(\tau)dx d\tau. \tag{1.4}$$

As $f(\tau)$ is a constant with respect to x, therefore the inner integral is simply $(t - \tau)f(\tau)$, and which further implies

$$ {}_0I_t^2 f(t) = \int_0^t (t-\tau)f(\tau)d\tau. \tag{1.5}$$

Similarly passing the function through a chain of 'n' integrals,

$$ {}_0I_t^n f(t) = \underbrace{\int \ldots \int_0^t}_{n} f(\tau)\underbrace{d\tau \ldots d\tau}_{n} = \int_0^t \frac{(t-\tau)^{n-1}}{(n-1)!} f(\tau)d\tau. \tag{1.6}$$

This completes the proof.

It is observed that:

- the above generalized integration is valid only for the integer points;
- it contains a factorial function;

However, one very sound mathematical observation comes into picture that the above definition contains the factorial function. So one question comes in front of researchers that can we "find a smooth curve that connects the points (n, n!) means factorial points" as shown in Fig. 1.1? Riemann-Liouville found that the generalized gamma function can provide a smooth curve connecting the factorial points. It is defined as

$$ \Gamma(z) = \int_0^\infty e^{-t} t^{z-1} dt, \tag{1.7}$$

which converges in the right half of the complex plane for $\mathbb{R}(z) > 0$.

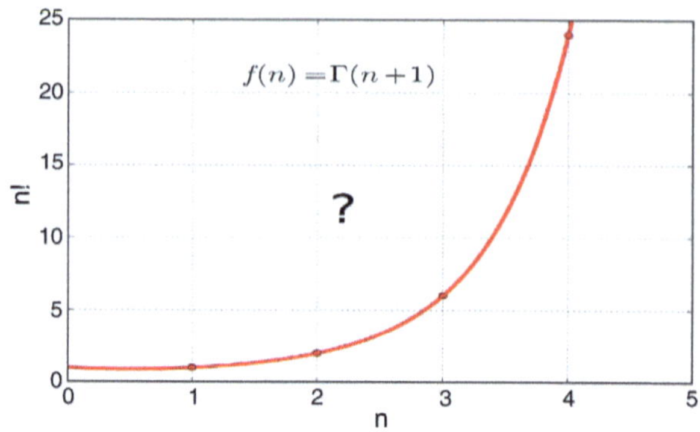

Fig. 1.1 Evolution of (n, n!)

Some of the important properties of gamma function are given below

•

$$
\begin{aligned}
\Gamma(z+1) &= \int_0^\infty e^{-t} t^{z+1-1} dt = \int_0^\infty e^{-t} t^z dt \\
&= [-e^{-t} t^z]_{t=0}^{t=\infty} + z \int_0^\infty e^{-t} t^{z-1} dt = z\,\Gamma(z). \qquad (1.8)
\end{aligned}
$$

- It has simple poles at $z = 0, -1, -2, -3, \ldots$, meaning is that the Gamma function asymptotically approaches infinity and is discontinuous at those negative integer values.
- **Fractional Differentiation**:
  Replacing the factorial $(n-1)!$ with Euler's continuous gamma function $\Gamma(n)$, which satisfies $(n-1)! = \Gamma(n)$ for $n \in \mathbb{R}$, one obtains a definition of a non-integer order integral, i.e.

Definition 1.2 The Riemann-Liouville (RL) definition of the αth-order fractional integration is given by

$$
{}_0I_t^\alpha f(t) := \frac{1}{\Gamma(\alpha)} \int_0^t \frac{f(\tau)}{(t-\tau)^{1-\alpha}} d\tau, \qquad (1.9)
$$

where α is in the set of positive real numbers.

Fractional Differentiation:
Definition of fractional order derivative operator is more straightforward. It is based on the knowledge of fractional order integral and the integer order differentiation. There are two possible ways to define it

- do enough times fractional integration firstly, then take sufficient number of integer differentiation which is known as Riemann-Liouville (RL) derivative,
- another way is just reverse of the above i.e. do enough times integer order differentiation firstly, then take sufficient number of fractional integration which is known as Caputo derivative.

Based on the first logic Riemann Liouville (RL) definition for the fractional-order derivative is given as

Definition 1.3 The Riemann Liouville definition for the fractional-order derivative of order $\alpha \in R^+$ and $m-1 < \alpha < m$, $m \in \mathbb{N}$ has the following form

$$ {}_0^{RL}D_t^{\alpha} f(t) := D^m D^{-(m-\alpha)} f(t) = \frac{d^m}{dt^m}\left[\frac{1}{\Gamma(m-\alpha)}\int_0^t \frac{f(\tau)}{(t-\tau)^{\alpha-m+1}}d\tau\right]. \tag{1.10}$$

An alternative definition for the fractional-order derivative was introduced by Caputo as

Definition 1.4 The Caputo definition for the fractional-order derivative of order $\alpha \in R^+$ and $m-1 < \alpha < m$, $m \in \mathbb{N}$ has the following form

$$ {}_0^{C}D_t^{\alpha} f(t) := D^{-(m-\alpha)} D^m f(t) = \frac{1}{\Gamma(m-\alpha)}\int_0^t \frac{f^m(\tau)}{(t-\tau)^{\alpha-m+1}}d\tau. \tag{1.11}$$

Above logic is supported by the following mathematical facts:

Remark 1.5 Denoting the derivative operator of order $n \in \mathbb{N}$ by D^n, and the identity operator by $\mathbb{I}$, we can verify that

$$D^n D^{-n} = \mathbb{I},\ \ D^{-n}D^n \neq \mathbb{I},\ \ n \in \mathbb{N},\ \ D^{-n}D^n f(t) = f(t) - \sum_{k=0}^{n-1} f^k(0^+)\frac{t^k}{k!}. \tag{1.12}$$

where $f^{(k)}(.)$ is the kth-order derivative of the function $f(.)$.

1.2.1 *Fractional Order Derivative Based on First Principle of Differentiation*

Suppose that some function $f(t)$ is sufficiently smooth. Then using the first principle of differentiation one can define

$$\dot{f}(t) = \frac{df(t)}{dt} = \lim_{h\to 0} \frac{f(t) - f(t-h)}{h}. \tag{1.13}$$

Differentiating $f(t)$ again

$$\begin{aligned}\ddot{f}(t) = \frac{d^2 f(t)}{dt^2} &= \lim_{h\to 0} \frac{\dot{f}(t) - \dot{f}(t-h)}{h} \\ &= \lim_{h\to 0} \frac{1}{h}\left[\frac{f(t) - f(t-h)}{h} - \frac{f(t-h) - f(t-2h)}{h}\right] \\ &= \lim_{h\to 0} \frac{f(t) - 2f(t-h) + f(t-2h)}{h^2}.\end{aligned} \tag{1.14}$$

Similarly, third derivative of $f(t)$ is given as

$$f^{(3)}(t) = \lim_{h\to 0} \frac{f(t) - 3f(t-h) + 3f(t-2h) - f(t-3h)}{h^3}. \tag{1.15}$$

Generalizing it for nth order as

$$f^{(n)}(t) = \lim_{h\to 0} \frac{1}{h^n} \sum_{r=0}^{n} (-1)^r \binom{n}{r} f(t - rh), \tag{1.16}$$

where

$$\binom{n}{r} = \frac{n(n-1)(n-2)\cdots(n-r+1)}{r!}. \tag{1.17}$$

Equation (1.16) can be generalized for arbitrary integer order α

$$f^{(\alpha)}(t) = \lim_{h\to 0} \frac{1}{h^\alpha} \sum_{r=0}^{n} (-1)^r \binom{\alpha}{r} f(t - rh). \tag{1.18}$$

Definition 1.6 The Grünwald-Letnikov derivative is based on the backward difference generalization and is given as

$$f^{(\alpha)}(t)|_{t=kh} = \Delta^\alpha(k) = \lim_{h\to 0} \frac{1}{h^\alpha} \sum_{r=0}^{k} (-1)^r \binom{\alpha}{r} f(kh - rh). \tag{1.19}$$

1.3 Why Fractional Order?

This is the one of the most important questions in front of researchers and academician, that what is the need of fractional order calculus if the integer order is enough for us? When we started the study on fractional order after discussion with one of our college friends, the same question come into our mind also. After looking into the literature towards we are able to find one important work which is presented by Blas M. Vinagre and YangQuan Chen [5], 41st IEEE CDC-2002 tutorial workshop. This gives lots of motivation to us, to work and try to bring more researchers, academicians and industrial people towards this area, particularly fractional order system in the science and control engineering disciple.

Before moving towards fractional order modeling and control a lot of restrictions of classical calculus is pointed out by Blas M. Vinagre and YangQuan Chen, starting from the following famous quotes given by the *Procrustes Bed : "all people must fit the same bed, if tall, cut the legs and if short, stretch the leg" and second one by the Purloined Letter : "same methods must be always successfully applicable, if no solution is found, there is no solution"*. If these two cases are not obvious, then one has to look towards a new paradigm and revisit the origins of classical one. One can see the same history behind the fractional order modeling and control over classical (integer) order.

- During the time domain modeling $g(t)$ is approximated using a linear combination of exponentials as shown in Fig. 1.2. Why not using a function, just one, being exponentials, real or complex, particular cases?
- During the frequency domain modeling $H(jw)$ is a rational function of s as shown in Fig. 1.3? Why not a rational function of $s^{0.4}$ in order to fit the slope with only a factor?
- Future evolution of the most of the practical system is dependent on the past value. Most important question, have all the past values, the same weight as shown in Fig. 1.4 or have different weights as shown in Fig. 1.5?
- Basic control actions PID, why not to extend in a continuous way as shown in Fig. 1.6 [6]?

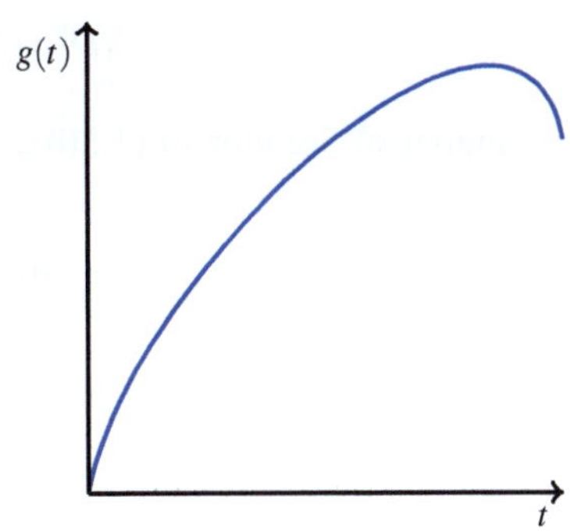

Fig. 1.2 Time domain modeling

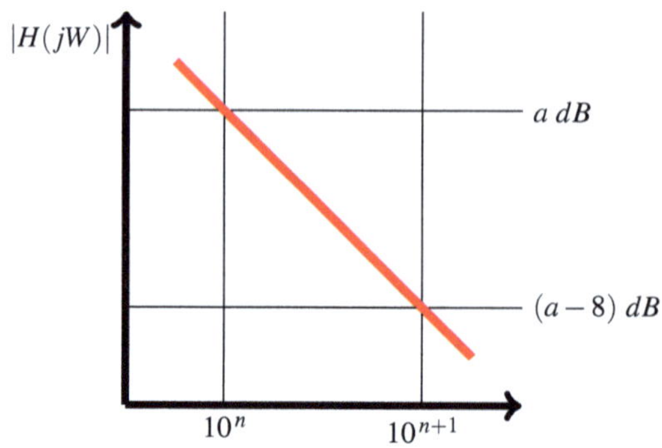

Fig. 1.3 Frequency domain modeling

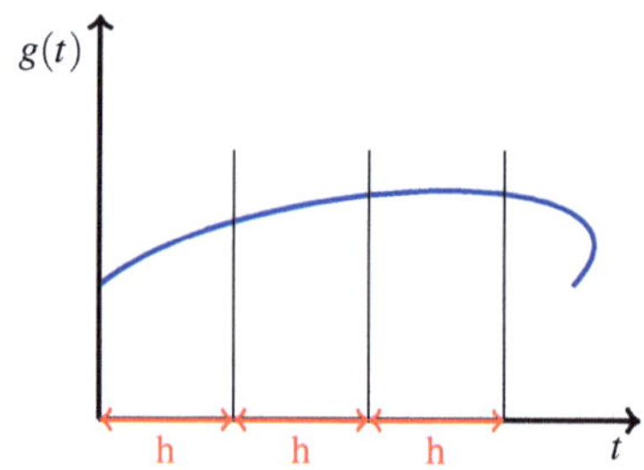

Fig. 1.4 Uniform time

- Reference system for control: why not to use a different and more robust (to gain and load changes) reference system for the design specifications as shown in Fig. 1.7 [7]?
- Additional attractive feature of fractional-order over integer-order system is in terms of stability. For example [8] comparing the following two systems with initial condition $x(0)$ for $0 < \mu < 1$

$$\dot{x}(t) = \mu t^{\mu-1}, \tag{1.20}$$

$${}_0^C D_t^{\alpha} x(t) = \mu t^{\mu-1}, \quad 0 < \alpha < 1. \tag{1.21}$$

Analytical Solution of (1.20) and (1.21) is found as

$$x(t) = t^{\mu} + x(0), \quad x(t) = \frac{\mu \Gamma(\mu) t^{\mu+\alpha-1}}{\Gamma(\mu+\alpha)} + x(0). \tag{1.22}$$

Obviously, the integer-order system (1.20) is unstable for any $\mu \in (0\ 1)$. However, the fractional dynamic system (1.21) is stable as $0 < \mu \le (1-\alpha)$, which implies that the fractional-order system may have additional feature over the integer-order system.

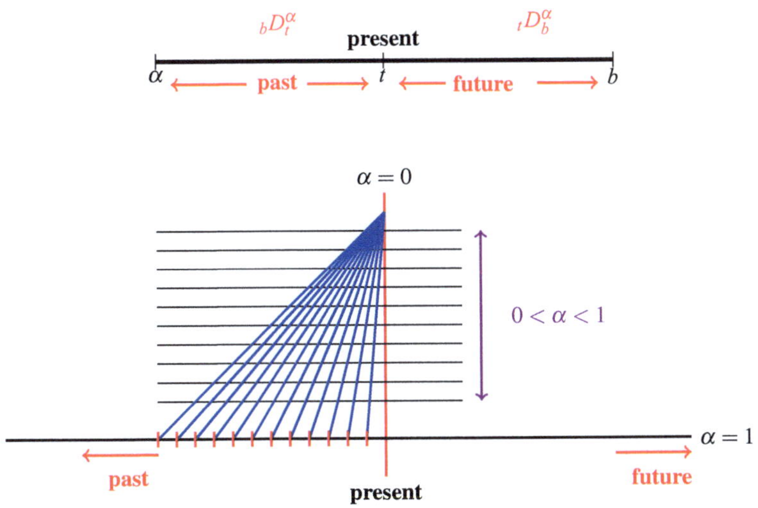

Fig. 1.5 Weighted time

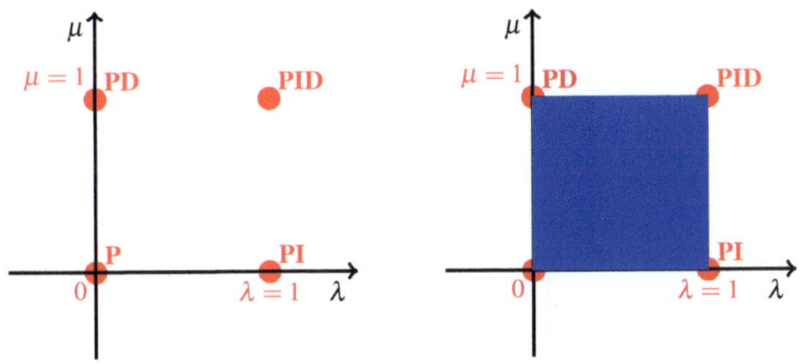

Fig. 1.6 Classical PID and Fractional PID

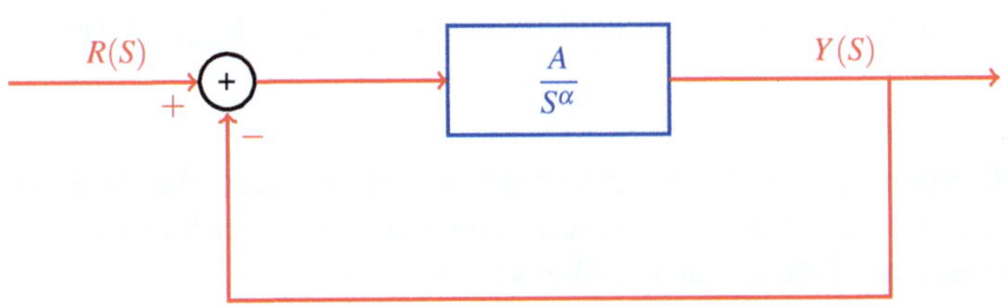

Fig. 1.7 Reference systems for control

- Some functions which is not differentiable in classical sense are found to be differentiable in fractional (RL) sense

Example Consider the following function [9]

$$x(t) = \begin{cases} 1 - t & \text{if } 0 < t \leq 1 \\ t - 1 & \text{if } 1 < t \leq 1 + \varepsilon. \end{cases} \tag{1.23}$$

The derivative (right) of $x(t)$ exists in the classical sense at $t = 0$, but doesn't exist derivative in the same sense at $t = 1$. By simple calculation, one can write

$${}_0^{RL}D_t^{\alpha} x(t) = \begin{cases} \dfrac{t^{-\alpha}}{\Gamma(1-\alpha)} - \dfrac{t^{1-\alpha}}{\Gamma(2-\alpha)} & \text{if } 0 < t \leq 1 \\ \dfrac{t^{-\alpha}}{\Gamma(1-\alpha)} - \dfrac{t^{1-\alpha} - 2(t-1)^{1-\alpha}}{\Gamma(2-\alpha)} & \text{if } 1 < t \leq 1 + \varepsilon, \end{cases} \tag{1.24}$$

where $\alpha \in (0, 1)$. From the above example, $x'(1)$ doesn't exist but RL derivative exists at $t = 1$.

Example Consider a discontinuous function which has a removal discontinuity point [10]

$$x(t) = \begin{cases} t & \text{if } 0 < t < 1 \\ 0 & \text{if } t = 1 \\ 2 - t & \text{if } 1 < t < 2. \end{cases} \tag{1.25}$$

Then, if $0 < \alpha < 1$, we get

$${}_0^{RL}D_t^{\alpha} x(t) = \begin{cases} \dfrac{t^{1-\alpha}}{\Gamma(2-\alpha)} & \text{if } 0 < t < 1 \\ \dfrac{t^{1-\alpha} - 2(t-1)^{1-\alpha}}{\Gamma(2-\alpha)} & \text{if } 1 < t < 2. \end{cases} \tag{1.26}$$

Therefore, $\lim\limits_{t \to 1^-} {}_0^{RL}D_t^{\alpha} x(t) = \lim\limits_{t \to 1^+} {}_0^{RL}D_t^{\alpha} x(t) = \frac{1}{\Gamma(2-\alpha)}$. Then ${}_0^{R}LD_t^{\alpha} x(t)$ exists at $t = 1$.

Example Riemann-Liouville derivative of the Weierstrass function (continuous everywhere but differentiable nowhere in classical sense) as shown in Fig. 1.8. One form of Weierstrass function is as follows [11]

$$W(t) = \sum_{j \geq 1} \lambda^{-\mu j} \sin(\lambda^j t), \quad (0 < \mu < 1, \lambda > 1). \tag{1.27}$$

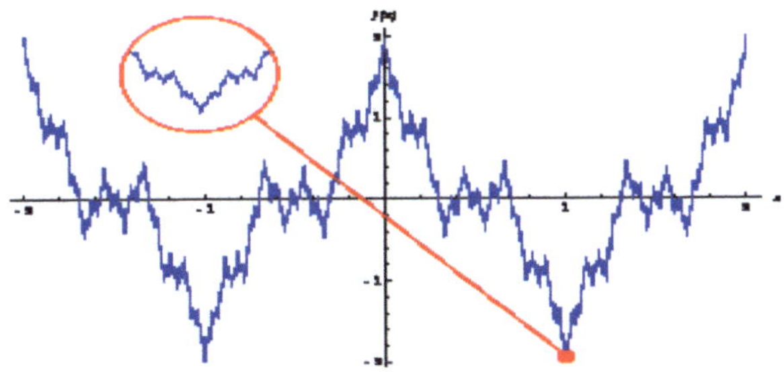

Fig. 1.8 Weierstrass function

Then one can get

$$
\begin{aligned}
{}_0^{RL}D_t^{-\alpha}W(t) &= \sum_{j\geq 1}\lambda^{-\mu j}S_t(\alpha,\lambda^j),\ \ 0<\alpha+\mu<1 \\
{}_0^{RL}D_t^{\beta}W(t) &= \sum_{j\geq 1}\lambda^{(1-\mu)j}C_t(1-\beta,\lambda^j),\ \ 0<\beta<\mu<1,
\end{aligned}
\tag{1.28}
$$

where

$$
\begin{aligned}
{}_0^{RL}D_t^{-\alpha}\text{sin}at &= \frac{1}{\Gamma}\int_0^t x^{\alpha-1}\text{sin}a(t-x)dx := S_t(\alpha,a) \\
{}_0^{RL}D_t^{-\alpha}\text{cos}at &= \frac{1}{\Gamma}\int_0^t x^{\alpha-1}\text{cos}a(t-x)dx := C_t(\alpha,a).
\end{aligned}
\tag{1.29}
$$

These results show that the αth order Riemann-Liouville integral and the βth order Riemann-Liouville derivative of the Weierstrass function exist for $0<\alpha+\mu<1$, $0<\beta<\mu<1$ correspondingly.

These all above facts, motivate researchers to look into the more generalized calculus for modeling and control of various physical and natural phenomenon.

1.4 Laplace Transform of Fractional Order Operators

The Laplace transform is a widely accepted integral transform in mathematics with many applications in physics and engineering. It is named after Pierre-Simon Laplace, who introduced the transform in his work on probability theory. It is a linear operator of a function $f(t)$ with a real argument $t(t\geq 0)$ that transforms $f(t)$ to a function $F(s)$ with complex argument s, given by the following integral relation

$$F(s) = \int_0^{\infty} f(t)e^{-st}dt. \tag{1.30}$$

The Laplace transformation is bijective for the majority of practical uses. This transformation has the most useful property that many relationships and operations over the original $f(t)$ correspond to simpler relationships and operations over its image $F(s)$. It also plays an important role in fractional order calculus same as integer order calculus. Some of the most frequently useful Laplace transformations of fractional order operators are discussed here [12].

Theorem 1.7 *The Laplace transformation of Riemann-Liouville integration is given as*

$$\mathscr{L}\left[{}_0I_t^{\alpha} f(t)\right] = s^{-\alpha} F(s), \tag{1.31}$$

and of RL derivative as

$$\mathscr{L}\left[{}_0D_t^{\alpha} f(t)\right] = s^{\alpha} F(s) - \sum_{k=0}^{m-1} s^k {}_0D_t^{\alpha-k-1} f(0), \tag{1.32}$$

where $\alpha \in R^+$ *and* $m - 1 < \alpha < m$, $m \in \mathbb{N}$.

Proof Taking Laplace transform action both of sides of fractional order integration, we get

$$\mathscr{L}\left[{}_0I_t^{\alpha} f(t)\right] = \mathscr{L}\left[\frac{1}{\Gamma(\alpha)} \int_0^t \frac{f(\tau)}{(t-\tau)^{1-\alpha}} d\tau\right]. \tag{1.33}$$

Following convolution theorem is used to simplify the above equation

$$\mathscr{L}\left[\int_0^t f(t-\tau)g(\tau)d\tau\right] = \mathscr{L}[f(t)]\mathscr{L}[g(t)]. \tag{1.34}$$

Using (1.33) and (1.34), one can write

$$\begin{aligned}
\mathscr{L}\left[{}_0I_t^{\alpha} f(t)\right] &= \frac{1}{\Gamma(\alpha)} \mathscr{L}\left[t^{\alpha-1}\right] \mathscr{L}[f(t)] \\
&= \frac{1}{\Gamma(\alpha)} \frac{\Gamma(\alpha)}{s^{\alpha}} \mathscr{L}[f(t)] \\
&= s^{-\alpha} \mathscr{L}[f(t)] \\
&= s^{-\alpha} F(s).
\end{aligned} \tag{1.35}$$

To obtain Laplace transform of Riemann-Liouville derivative, consider the following result,

For $\alpha < 0$, using Riemann-Liouville integral one can write

$$\mathscr{L}\left[{}_0D_t^{\alpha} f(t)\right] = \mathscr{L}\left[\frac{1}{\Gamma(-\alpha)}\int_0^t \frac{f(\tau)}{(t-\tau)^{1+\alpha}}d\tau\right]. \tag{1.36}$$

Further using convolution theorem

$$\begin{aligned}\mathscr{L}\left[{}_0D_t^{\alpha} f(t)\right] &= \frac{1}{\Gamma(-\alpha)}\mathscr{L}\left[t^{-\alpha-1}\right]\mathscr{L}\left[f(t)\right],\\ &= \frac{1}{\Gamma(-\alpha)}\frac{\Gamma(-\alpha)}{s^{-\alpha}}\mathscr{L}\left[f(t)\right],\\ &= s^{\alpha}\mathscr{L}\left[f(t)\right].\end{aligned} \tag{1.37}$$

Also, for the nth integer order derivative, Laplace transform of the same is expressed as

$$\mathscr{L}\left[{}_0D_t^{n} f(t)\right] = s^n F(s) - \sum_{k=0}^{n-1} s^k \frac{d^{n-k-1}f(0)}{dt^{n-k-1}}. \tag{1.38}$$

Laplace transform of Riemann-Liouville derivative is given as

$$\begin{aligned}\mathscr{L}\left[{}_0D_t^{\alpha} f(t)\right] &= \mathscr{L}\left[D^m{}_0D_t^{\alpha-m} f(t)\right],\\ &= s^m s^{\alpha-m}F(s) - \sum_{k=0}^{m-1} s^k D^{m-k-1}{}_0D_t^{\alpha-m} f(0).\end{aligned} \tag{1.39}$$

This completes the proof.

The Laplace transformation of another popular fractional order derivative known as Caputo's derivative is considered next.

Theorem 1.8 *The Laplace transformation of Caputo derivative is given as*

$$\mathscr{L}\left[{}_0^CD_t^{\alpha} f(t)\right] = s^{\alpha}F(s) - \sum_{k=0}^{m-1} s^{\alpha-k-1}D^k f(0), \tag{1.40}$$

where $\alpha \in R^+$ *and* $m-1 < \alpha < m$, $m \in \mathbb{N}$.

Proof Using Caputo's definition one can write

$$\mathscr{L}\left[{}_0^C D_t^\alpha f(t)\right] = \mathscr{L}\left[{}_0 D_t^{\alpha-m} D^m f(t)\right],$$
$$= s^{\alpha-m}\left[s^m F(s) - \sum_{l=0}^{m-1} s^l D^{m-l-1} f(0)\right]. \tag{1.41}$$

Substituting $k = m - 1 - l$ in (1.41)

$$\mathscr{L}\left[{}_0^C D_t^\alpha f(t)\right] = s^\alpha F(s) - \sum_{k=0}^{m-1} s^{\alpha-k-1} D^k f(0).$$

Thus, the Laplace transform is verified.

Remark 1.9 It can be noticed that Laplace transformation of the RL derivative requires information of initial conditions of the function i.e. ${}_0D_t^{\alpha-k-1} f(0)$(see Eq. (1.32)), This is found to be difficult to measure in term of physical quantities. However in the Caputo's definition initial condition is same as the integer derivatives of f, which has well known physical interpretation. This gives a strong support for acceptance of Caputo's derivative to researchers and practicing engineers.

Consider a question, does there exist an equivalent integer order LTI/LTV system to a fractional order system? If yes, then fractional order systems can be regarded as a part of integer order, or if answer is no, then fractional order systems can not be ignored and the research of FOS is magnificent innovative. This equivalence has been discussed in the next section.

1.5 Equivalence Between Fractional Order System (FOS) and Integer Order System (IOS)

Consider the following time invariant fractional order system

$$D^\alpha x(t) = Ax(t). \tag{1.42}$$

Is the system (1.42), equivalent to any of the following systems [13]?

- linear integer order time invariant system

$$\dot{x}(t) = A_1 x(t), \tag{1.43}$$

- integer order time varying system

$$\dot{x}(t) = A_2(t)x(t), \tag{1.44}$$

- or integral system represented as

$$\dot{x}(t) = \frac{A}{\Gamma(\alpha)} \int_0^t (t-\tau)^{(\alpha-1)} x'(\tau) d\tau + \frac{Ax(0)t_+^{\alpha-1}}{\Gamma(\alpha)}. \tag{1.45}$$

For checking the above equivalence following Lemma and Theorem are required.

Lemma 1.10 *The Laplace transform of* $\frac{t_+^{-\frac{1}{2}}}{\Gamma(\frac{1}{2})}$ *is:*

$$L\left(\frac{t_+^{-\frac{1}{2}}}{\Gamma(\frac{1}{2})}\right) = s^{-\frac{1}{2}}. \tag{1.46}$$

Theorem 1.11 *For* $0 < \alpha < 1$, *system* (1.42) is equivalent to (1.45).

Proof Using Laplace transform for (1.42) and taking into account the Caputo's definition for fractional-order derivative, we obtain

$$s^{\alpha} X(s) - s^{\alpha-1} x(0) = AX(s). \tag{1.47}$$

Pre multiplying by $s^{1-\alpha}$, yields

$$\begin{aligned} sX(s) - x(0) &= As^{1-\alpha} X(s) \\ &= A(s^{1-\alpha} X(s) - s^{-\alpha} x(0)) + As^{-\alpha} x(0). \end{aligned} \tag{1.48}$$

Substituting $\alpha = \frac{1}{2}$ and taking inverse Laplace transform, we have

$$\dot{x}(t) = A^2 x(t) + \frac{Ax(0)t_+^{-\frac{1}{2}}}{\Gamma(\frac{1}{2})}. \tag{1.49}$$

If we denote $B = Ax(0)$, $u(t) = \frac{t_+^{-\frac{1}{2}}}{\Gamma(\frac{1}{2})}$, then (1.49) changes as follows

$$\dot{x} = A^2 x(t) + Bu(t) \tag{1.50}$$

Clearly the equivalent representation as shown in (1.45) is exactly identical to the solutions (1.42) obtained as in (1.50). This completes the proof of Theorem.

Remark 1.12 As $\int_0^\infty u(\tau)d\tau = \int_0^\infty \frac{\tau_+^{-1/2}}{\Gamma(1/2)} d\tau = \infty$, it can be conclude that $u(t)$ is not a proper function, i.e. in reality, there does not exist an input function $u(t)$ with limit power satisfying integer order system (1.50) which is equivalent to fractional order system (1.42).

Remark 1.13 For real world system model, it is inconceivable that the initial input and the amount of input $u(t)$ is infinity. It might happen only at the beginning of cosmic. But using equivalence between (1.42) and (1.50), one can discuss some properties like stability of FOS (fractional order stability) and Mittag-Leffler function.

1.6 Physical Realization of the Fractional Derivatives and Integral

After successful development of sufficient mathematical background on fractional order (non-integer) operators, most important question in front of the researchers that how to realize it at least theoretically. If it is possible then it might be possible for people to use it to represent practical problems using some approximation. In this direction most successful effort was made by Oldham and Spanier [14] and also by Oldham and Zoski [15]. They suggested two different electronic circuit that can perform various orders of fractional operators. The circuit given by Oldham and Spanier, for realization of fractional operator is discussed below.

1.6.1 Chain Fractance

It is obvious from the definition of the fractional operators that generalization of differentiation is motivated by fractional order integration. In other way one can interpret that integration is the basic element for the fractional order operators. It is also well known fact that one can realize the classical integrator by using the capacitor circuit, if its initial voltage $v(0)$ is zero, which is shown in Fig. 1.9. Let us

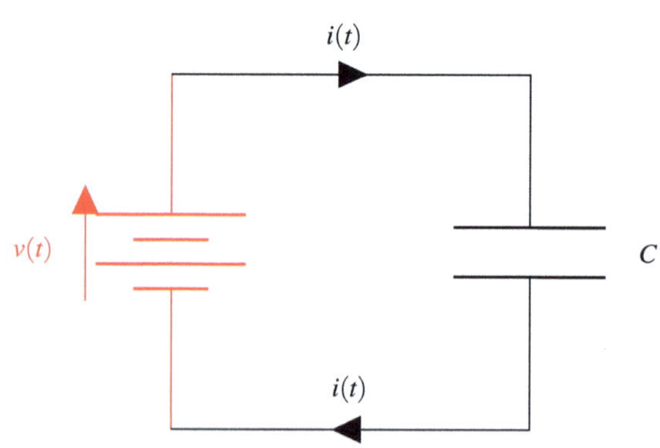

Fig. 1.9 A basic integrating circuit

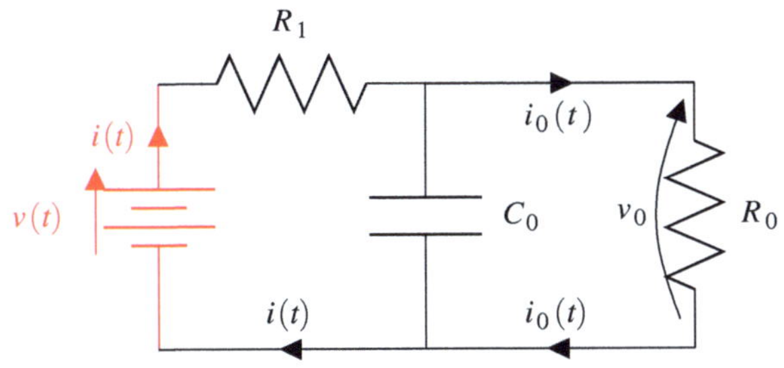

Fig. 1.10 Three-component RC circuit

assume that applied voltage across the capacitor is $v(t)$, the current flowing through the capacitor is $i(t)$ and C is the capacitance of circuit. Then one can write

$$v(t) = \frac{1}{C}\frac{d^{-1}}{dt^{-1}}i(t), \tag{1.51}$$

provided the initial voltage across the capacitor $v(0) = 0$.

Now, for constructing the circuit for the fractional order integrator Oldham and Spanier [14] take the following simple three-component circuit as shown in Fig. 1.10. The current through the resistor R_1 is $i(t)$ and it also contains the extra elements, which is parallel combination of the capacitor C_0 and the resistor R_0. Let us assume that initial condition of current and voltage both are zero at the time $t = 0$. A nonzero voltage $v(t)$ is applied to the circuit (Fig. 1.10) after $t \geq 0$ and assume that current through R_0 is i_0 and voltage across the capacitor is $v_0(t)$. Therefore, one can write

$$i_0(t) = \frac{v_0(t)}{R_0}, \tag{1.52}$$

and

$$i(t) - i_0(t) = C_0\frac{dv_0(t)}{dt}. \tag{1.53}$$

Using Ohm's law it is obvious that

$$i(t) = \frac{v(t) - v_0(t)}{R_1}. \tag{1.54}$$

Eliminating $v_0(t)$ and $i_0(t)$ from (1.52–1.54)

$$(R_0 + R_1)i(t) + R_0R_1C_0\frac{di(t)}{dt} = v(t) + R_0C_0\frac{dv(t)}{dt}. \tag{1.55}$$

Taking Laplace transform action of (1.55), one can further write

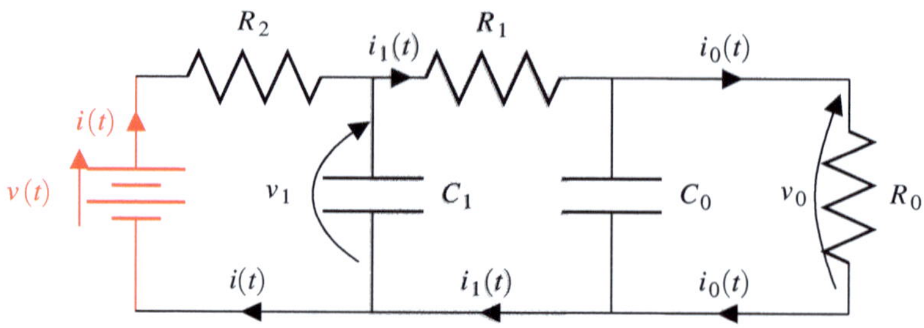

Fig. 1.11 Further generalization of RC circuit

$$(R_0 + R_1 + R_0 R_1 C_0 s) I(s) - R_0 R_1 C_0 i(0) = (1 + R_0 C_0 s) V(s) - R_0 C_0 v(0), \tag{1.56}$$

where $I(s) = \mathscr{L}\{i(t)\}$ and $V(s) = \mathscr{L}\{v(t)\}$. After substituting the initial condition $i(0) = v(0) = 0$, (1.56) can be written as

$$\frac{V(s)}{R_1 I(s)} = 1 + \frac{\frac{1}{R_1 C_0}}{s + \frac{1}{R_0 C_0}}. \tag{1.57}$$

For the further generalization by Oldham and Spanier [14], a series resistor R_2 and a shunt capacitor C_1 has been added to the circuit as shown in Fig. 1.11. The current-voltage relation of this circuit can be written as

$$i(t) = \frac{v(t) - v_1(t)}{R_2} \quad \text{and} \quad i(t) - i_1(t) = C_1 \frac{dv_1(t)}{dt},$$

where $i(t)$ and $i_1(t)$ are the currents as shown in Fig. 1.11. In a similar way as done for the circuit in Fig. 1.10, one can write

$$\frac{V(s)}{R_2 I(s)} = 1 + \frac{\frac{1}{R_2 C_1}}{s + \frac{I_1(s)}{C_1 V_1(s)}}. \tag{1.58}$$

Similarly, one can also write

$$\frac{V_1(s)}{R_1 I(s)} = 1 + \frac{\frac{1}{R_1 C_0}}{s + \frac{1}{R_0 C_0}}. \tag{1.59}$$

Consider the following definitions

$$f_0 := \frac{1}{R_0 C_0}, \quad f_1 := \frac{1}{R_1 C_0}, \quad f_2 := \frac{1}{R_1 C_1}, \quad f_3 := \frac{1}{R_2 C_1}. \tag{1.60}$$

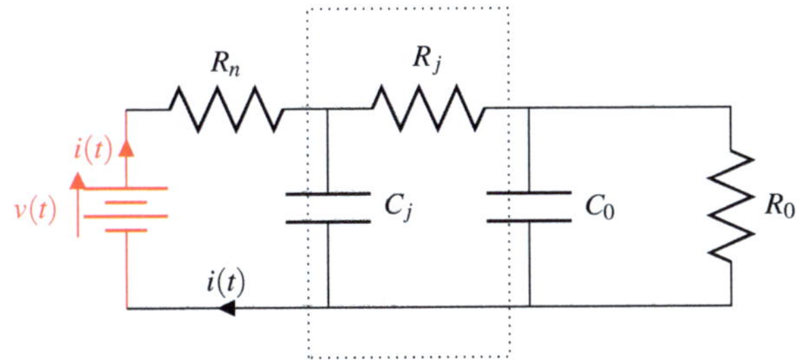

Fig. 1.12 Generalized RC circuit where $j = 1, 2, \ldots, n-1$

Combining (1.58–1.60), one can write

$$\frac{V(s)}{R_2 I(s)} = 1 + \frac{f_3}{s + \frac{f_2}{1 + \frac{f_1}{s + f_0}}}. \tag{1.61}$$

For simplicity, (1.61) can be rewritten using the notation of continued functions as

$$\frac{V(s)}{R_2 I(s)} = 1 + \frac{f_3}{s+}\,\frac{f_2}{1+}\,\frac{f_1}{s+}\,\frac{f_0}{1}. \tag{1.62}$$

For the further generalization of (1.62) following circuit was constructed by Oldham and Spanier [14], as shown in Fig. 1.12. Using Fig. 1.12, one can write

$$\frac{V(s)}{R_n I(s)} = 1 + \frac{f_{2n-1}}{s+}\,\frac{f_{2n-2}}{1+}\,\frac{f_{2n-3}}{s+} \cdots \frac{f_2}{1+}\,\frac{f_1}{s+}\,\frac{f_0}{1}. \tag{1.63}$$

where $f_{2j} = \frac{1}{R_j C_j}$ and $f_{2j+1} = \frac{1}{R_{j+1} C_j}$ for $j = 0, 1, \ldots, n-1$. Defining $\frac{f_j}{s} := \eta_j$ and substituting η_j in (1.63), one can write

$$\frac{V(s)}{R_n I(s)} = 1 + \frac{\eta_{2n-1}}{1+}\,\frac{\eta_{2n-2}}{1+}\,\frac{\eta_{2n-3}}{1+} \cdots \frac{\eta_2}{1+}\,\frac{\eta_1}{1+}\,\frac{\eta_0}{1}. \tag{1.64}$$

Consider the following special case for the capacitors and resistors in Fig. 1.12

- $C_0 = C_1 = C_2 = \cdots = C_{n-1} := C$
- $R_0 = R_1 = R_2 = \cdots = R_{n-1} := R$ and $R_n := \dfrac{R}{2}$.

Thus

- $\eta_0 = \eta_1 = \eta_2 = \cdots = \eta_{2n-2} = \dfrac{1}{RCs} := \eta$
- $\eta_{2n-1} = \dfrac{2}{RCs} = 2\eta$.

Therefore, using (1.64) one can write

$$\frac{2V(s)}{RI(s)} = 1 + 2\frac{\eta}{1+}\frac{\eta}{1+}\frac{\eta}{1+}\cdots\frac{\eta}{1+}\frac{\eta}{1+}\frac{\eta}{1}. \tag{1.65}$$

where the continued fractional has $2n$ numeratorial $\eta's$. Further using the well known result of continued fractions given by Wall [16], one can show that

$$\frac{\eta}{1+}\frac{\eta}{1+}\frac{\eta}{1+}\cdots\frac{\eta}{1+}\frac{\eta}{1+}\frac{\eta}{1} = \frac{\sqrt{4\eta+1}}{1+\left[\frac{\sqrt{4\eta+1}-1}{\sqrt{4\eta+1}+1}\right]^{2n+1}} - \frac{\sqrt{4\eta+1}}{2} - \frac{1}{2}. \tag{1.66}$$

Using (1.64), (1.65) and dividing by $2\sqrt{\eta}$, we obtain

$$\frac{V(s)}{I(s)}\sqrt{\frac{Cs}{R}} = \sqrt{\frac{4\eta+1}{4\eta}}\left[\frac{\left[\sqrt{4\eta+1}+1\right]^{2n+1} - \left[\sqrt{4\eta+1}-1\right]^{2n+1}}{\left[\sqrt{4\eta+1}+1\right]^{2n+1} + \left[\sqrt{4\eta+1}-1\right]^{2n+1}}\right]. \tag{1.67}$$

For the larger value of n, the right side of (1.67) is very close to unity provided $6 \leq \eta \leq \frac{n^2}{6}$. In fact, for n in excess of about 10, the right hand side of (1.67) lies within 2 % of unity,

$$\frac{V(s)}{I(s)}\sqrt{\frac{Cs}{R}} \approx 1, \tag{1.68}$$

provided

$$6 \leq \eta \leq \frac{n^2}{6}. \tag{1.69}$$

In that case one can also write (1.68) as

$$V(s) \approx \sqrt{\frac{R}{Cs}}I(s) \text{ for } 6RC \leq \frac{1}{s} \leq \frac{n^2 RC}{6}. \tag{1.70}$$

Using inverse Laplace transform of (1.70),

$$v(t) \approx \sqrt{\frac{R}{C}}\frac{d^{-\frac{1}{2}}i(t)}{dt^{-\frac{1}{2}}} \text{ for } 6RC \leq \frac{1}{s} \leq \frac{n^2 RC}{6}. \tag{1.71}$$

Hence, using (1.70) one can conclude that the circuit shown in Fig. 1.12 is able to calculate the fractional order integration. However, the resultant integration is bounded by the same time limit, which can be adjusted by adjusting the value of capacitance and resistance. The other circuit performing fractional operations, given by Oldham and Zoski, is discussed in the next subsection.

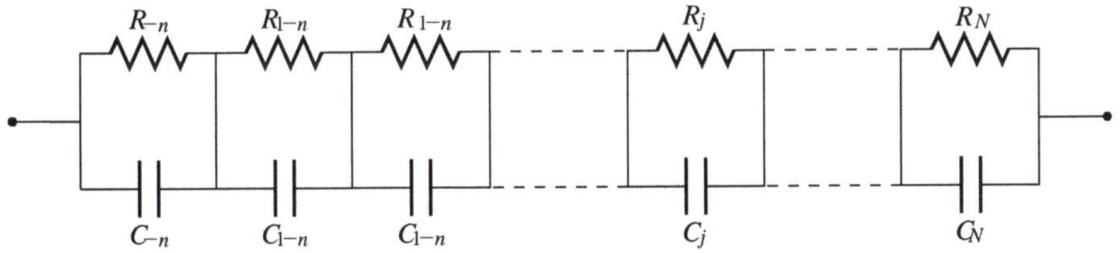

Fig. 1.13 Domino Ladder network

1.6.2 Domino Ladder

For the realization of fractional operators Oldham and Zoski [15] used *domino ladder* network that consists of a chain of resistors and a chain of capacitors, the two chains being interconnected at each node as shown in Fig. 1.13. It is assumed that the circuit is at rest, with all capacitors uncharged, prior to $t = 0$.

It is shown by Oldham and Zoski [15] that if the time dependent current $i(t)$ is passed through the network, a potential difference $V_j(t)$ across the jth pair of the components of domino ladder network, is related as

$$i(t) = \frac{V_j(t)}{R_j} + C_j \frac{d}{dt} V_j(t). \tag{1.72}$$

Here R_j and C_j represent jth component of resistor and capacitor respectively. Solving above equation using Laplace transformation, one can write

$$V_j(t) = \frac{1}{C_j} \int_0^t i(t-\tau) e^{\left(-\frac{\tau}{R_j C_j}\right)} d\tau. \tag{1.73}$$

The potential across the entire network for the common value of $i(t)$ is given as

$$V(t) = \int_0^t i(t-\tau) \sum_{j=-n}^{N} \frac{1}{C_j} e^{\left(-\frac{\tau}{R_j C_j}\right)} d\tau. \tag{1.74}$$

Further they changed the value of adjacent resistors and capacitors by constant factors $g > 1$ and $G > 1$ respectively and formed two geometric progressions as $R_j = g^{-j} R_0$ and $C_j = G^{-j} C_0$. Then the modified potential across the entire network for the common value of $i(t)$ is given as

$$V(t) = \frac{R_0^{\alpha}}{C_0^{1-\alpha}} \int_0^t \frac{i(t-\tau)}{\tau^{\alpha}} \sum_{j=-n}^{N} \left(\frac{\tau G^j g^j}{R_0 C_0} \right) e^{\left(-\frac{\tau G^j g^j}{R_0 C_0}\right)} d\tau, \tag{1.75}$$

where $\alpha = \frac{\ln(G)}{\ln(Gg)}$. This potential across the domino ladder network is further modified as (in the limiting case, when n and N both approach towards infinity and g and G both approach unity)

$$V(t) = \frac{\pi \operatorname{cosec}(\alpha\pi) R_0^{\alpha}}{\Gamma(1-\alpha)\ln(Gg)C_0^{1-\alpha}} \int_0^t \frac{i(t-\tau)}{\tau^{\alpha}} d\tau$$

$$= \frac{\pi \operatorname{cosec}(\alpha\pi) R_0^{\alpha}}{\ln(Gg)C_0^{1-\alpha}} \frac{d^{\alpha-1}}{dt^{\alpha-1}} i(t). \tag{1.76}$$

Using (1.76) one can clearly visualize the (non-integer) half order fractional operator for the particular case when $g = G$. This particular case implies that in domino ladder both resistors and capacitors have the same ratio. Further in more concise way one can explain that relation between potential difference and time varying current across the domino ladder network follows the non-integer integral.

The derivation (1.76) is not straight forward, so the following remark is included to derive the above relation.

Remark 1.14 Consider the following function

$$F(z) = e^{(\alpha z - e^{z})}, \tag{1.77}$$

where $\alpha > 1$. The above function has a maximum value at $z = ln\alpha$ and approaches $F = 0$ asymptotically as $z \to \pm\infty$. The area enclosed by this curve is finite and one can show it is equal to gamma function of α as follows

$$\int_{-\infty}^{\infty} F(z)dz = \Gamma(\alpha). \tag{1.78}$$

Using Riemann sum definition of integral, one can also write

$$lim \left[\delta \sum_{j=-n}^{N} (z_0 + \delta + j\delta) \right] = \int_{-\infty}^{\infty} F(z)dz = \Gamma(\alpha), \tag{1.79}$$

for any finite value of z_0, where "lim" signifies the simultaneous, independent limits taken as $\delta \to 0, n \to \infty$ and $N \to \infty$. Now substituting $z_0 = \frac{ln\tau}{GgR_0C_0}$ and $\delta = ln(Gg)$ in (1.79), one can get

$$ln(Gg) \sum_{j=-n}^{N} \left(\frac{\tau G^j g^j}{R_0 C_0} \right)^{\alpha} e^{-\left(\frac{\tau G^j g^j}{R_0 C_0}\right)} = \frac{\pi \operatorname{cosec}(\alpha\pi)}{\Gamma(1-\alpha)}. \tag{1.80}$$

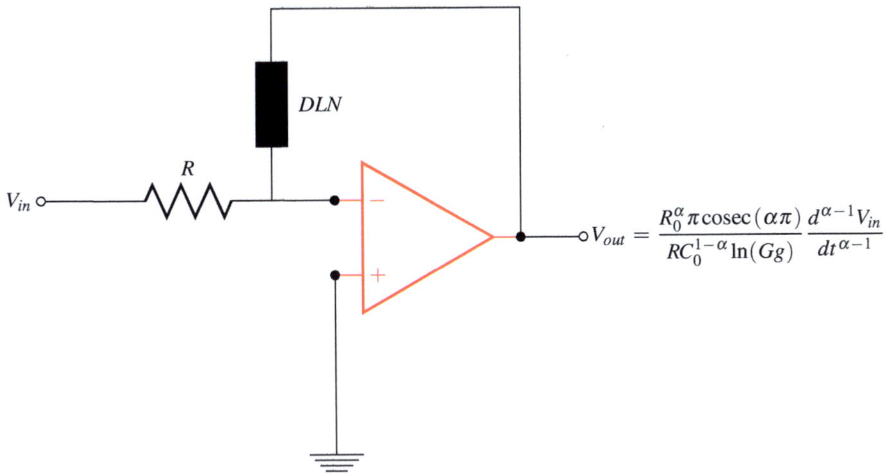

Fig. 1.14 Circuit for Fractional Operator $-4/5$

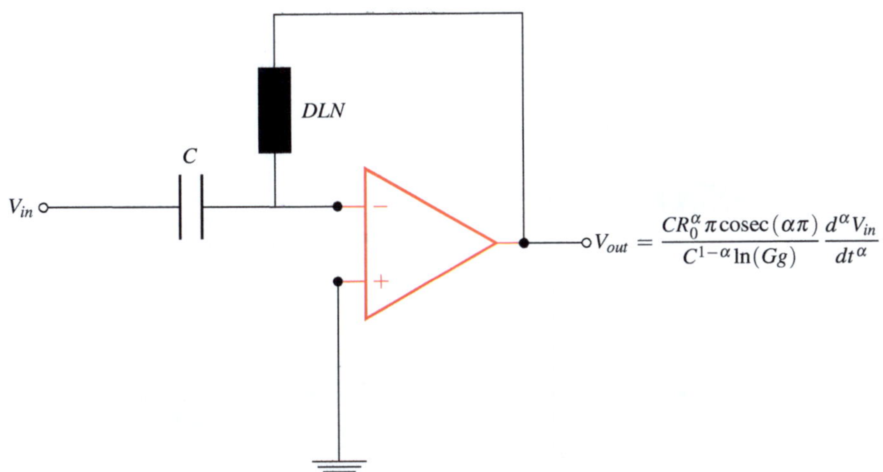

Fig. 1.15 Circuit for Fractional Operator $1/5$

Using domino ladder network (DLN) Oldham and Zoski [15] also constructed fractional order differentiator. For example, a ladder constructed for $\alpha = \frac{1}{5}$ can perform any of the following operations [15]

$$\frac{d^{-4/5}V(t)}{dt^{-4/5}}, \quad \frac{d^{1/5}V(t)}{dt^{1/5}}, \quad \frac{d^{-1/5}V(t)}{dt^{-1/5}}, \quad \frac{d^{4/5}V(t)}{dt^{4/5}}. \tag{1.81}$$

as shown in Figs. 1.14, 1.15, 1.16 and 1.17 respectively.

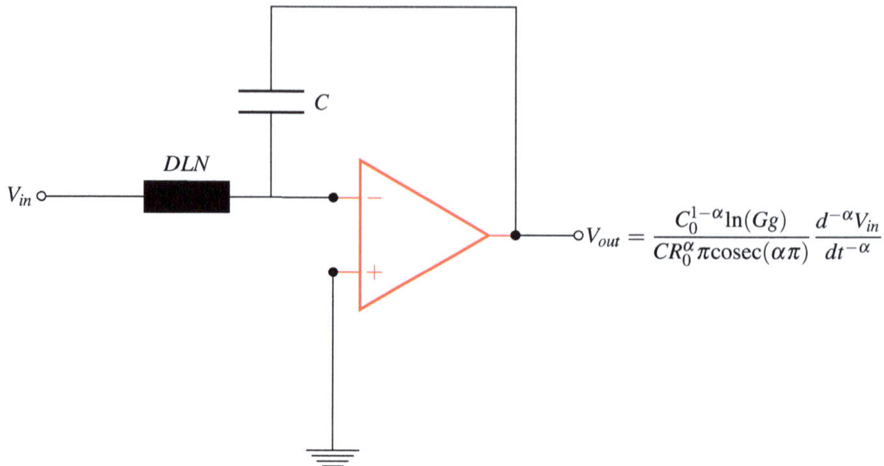

Fig. 1.16 Circuit for Fractional Operator $-1/5$

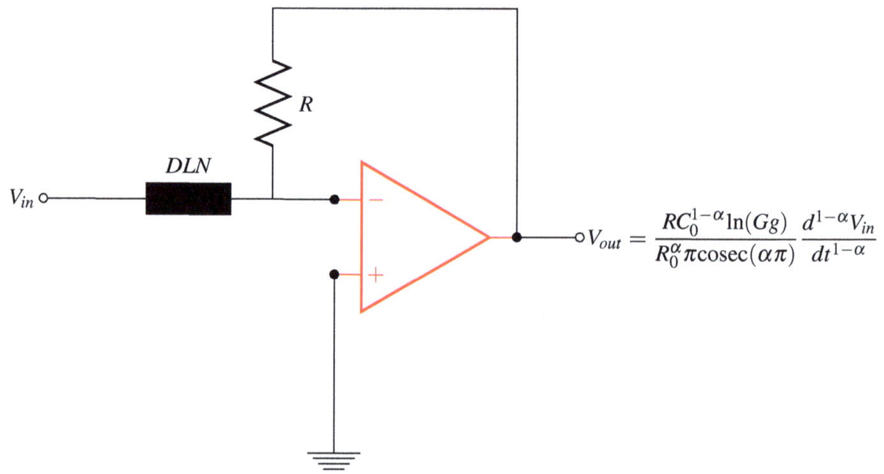

Fig. 1.17 Circuit for Fractional Operator $4/5$

1.7 Interpretation of Fractional Integral and Derivative

Fractional calculus provides a platform for the generalization of standard integer order integrals and derivatives to a non-integer order or even complex order. In this line of thought a wide range of physical phenomena may be captured. However, until recently, fractional calculus was an unknown mathematical tool for applied science, due to the the following reasons

- plethora of different definitions for fractional derivatives and integrals;
- absence of efficient computation tools;
- sufficiency of classical differential calculus for real world problem;
- and the most important fact that a simple interpretation for fractional derivative and integral is difficult.

Having these ideas in mind, several researchers proposed different approaches for the interpretation of fractional-order integrals and derivatives. Some of the most successful efforts are discussed in this Chapter [17–19].

1.7.1 Interpretation of Grünwald-Letnikov Derivative

Grünwald-Letnikov definition of a derivative of fractional order α of signal $f(t)$ is, given by

$$D^{\alpha}f(t) = \lim_{h \to 0} \frac{1}{h^{\alpha}} \sum_{k=0}^{\infty} \gamma(\alpha, k) f(t - kh). \tag{1.82}$$

$$\gamma(\alpha, k) = (-1)^k \frac{\Gamma(\alpha + 1)}{k! \Gamma(\alpha - k + 1)}, \tag{1.83}$$

where Γ and h are the gamma function and time increment respectively. One can also see the following relation from (1.83) for $0 < \alpha < 1$:

- $$\gamma(\alpha, 0) = 1, \tag{1.84}$$

  Probabilistic Interpretation: The *present* (i.e. $f(0)$) is seen in expression (1.82) with probability one.
- $$-\sum_{k=1}^{\infty} \gamma(\alpha, k) = 1. \tag{1.85}$$

  Probabilistic Interpretation: The *past/future* (i.e. $f(-h), f(-2h), \ldots$) is also captured with probability one. However, each sample $f(t)$ is weighted with given probability, that gets higher the closer we are to the *present*.

Observation: Expression $\sum_{k=1}^{\infty} \gamma(\alpha, k) f(t - kh)$ can be interpreted as the expected value $E(X)$ of the random variable X, such that $P(X = f(kh)) = |\gamma(\alpha, k)|, k = 1, 2, \ldots, 0 < \alpha < 1$ where P represents probability.
Geometrical Interpretation: The geometric interpretation of Grünwald-Letnikov derivative is the slope θ of a triangle composed by the initial value of function $f(0)$

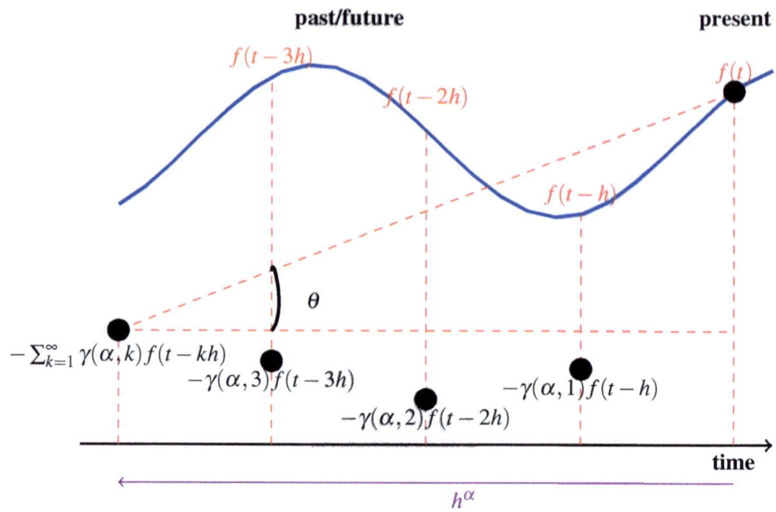

Fig. 1.18 Geometric interpretation of Grünwald-Letnikov derivative

and expected value $E(X)$ placed at location $t = h^{\alpha}$, that is the *present* sample of the signal $f(t)$ and the arithmetic average of the *past/future*. At the limiting situation of the increment $h \to 0$, the slope of above mentioned triangle θ reflects the Grünwald-Letnikov derivative (i.e. as $h \to 0$ the slope $\theta \to D^{\alpha} f(t)$, as shown in Fig. 1.18).

Remark 1.15 For $\alpha = 1$ (in the case of integer order) *past/future* has probability one just near the *present* sample and zero the rest of *past/future*, that corresponds to slope of tangent line at *present* sample, which reflects the same interpretation as integer order derivative. One can also conclude that integer order derivative corresponds to a deterministic perspective which is a particular case of more general fractional derivative.

1.7.2 Inhomogeneous (Cosmic) Time Scale Based Interpretation of Fractional Integral and Derivative

αth order Riemann-Liouville fractional integral is given as

$$ {}_0I_t^{\alpha} f(t) = \frac{1}{\Gamma(\alpha)} \int_0^t f(\tau)(t-\tau)^{\alpha-1} d\tau, \quad t \geq 0. \tag{1.86}$$

Equation (1.86) can be also expressed as

Fig. 1.19 Homogeneous time scale

Fig. 1.20 Nonhomogeneous time scale

$$ {}_0I_t^{\alpha} f(t) = \frac{1}{\Gamma(\alpha)} \int_0^t f(\tau) dg_t(\tau), \tag{1.87} $$

where

$$ g_t(\tau) = \frac{1}{\Gamma(\alpha+1)} \left\{ t^{\alpha} - (t-\tau)^{\alpha} \right\}. \tag{1.88} $$

One of the best known attempts of visual geometric interpretation of fractional integral and derivative was given by Podlubny [19], which is based on two kinds of time, the cosmic time $(g_t(\tau))$ and individual time τ (Fig. 1.18).

- **Homogeneous Time**: It corresponds to equably flowing time as shown in Fig. 1.19. The entire classical calculus (integer calculus) is based on it.
- **Cosmic Time**: It corresponds to non-equably flowing time as shown in Fig. 1.20. The change of scale of the cosmic time is described using the homogeneous time scale as a reference scale. In other words, one can interprets the homogeneous time scale as an ideal notion, which is necessary for developing mathematical models describing inhomogeneous cosmic time and its change.

Scaling property of cosmic time function $(g_t(\tau))$ is given as

$$ g_{t_1}(\tau_1) = g_{kt}(k\tau) = k^{\alpha}(g_t(\tau)), \tag{1.89} $$

where $t_1 = kt$ and $\tau_1 = k\tau,\ k > 0$.

Let us now consider a three-dimensional space with co-ordinate (τ, g, f) and plot the function $g_t(\tau)$ for $0 \le \tau \le t$ on the plane (τ, g) as shown in Fig. 1.21. Along the obtained curve, by varying the height $f(\tau)$ one can build a **fence**. Finally **fence** can be projected onto two surfaces

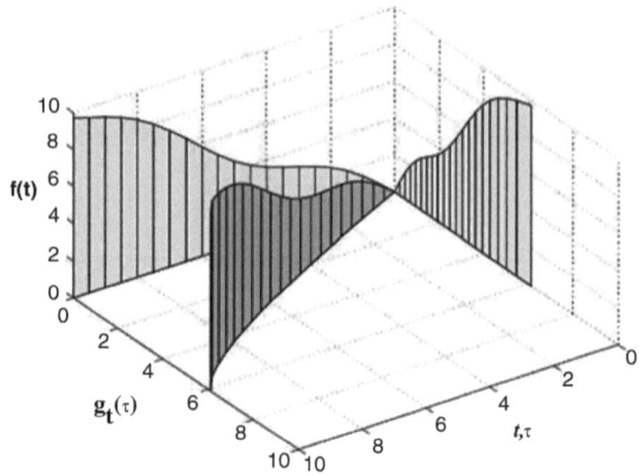

Fig. 1.21 Fence

- the first of them on the wall (τ, f) and interpret it as "area under the curve $f(\tau)$"

$$ {}_0I_t^1 f(t) = \frac{1}{\Gamma(\alpha)} \int_0^t f(\tau)d(\tau), \tag{1.90}$$

  which is standard geometrical interpretation of first order classical integral.
- the second one is the area of the projection of the same **fence** on the wall (g, f), which provides the geometrical interpretation of fractional integral same as in (1.87).

Finally Podlubny [19] argued the physical significance of cosmic time based on real physical phenomenon. He discussed that when a object is moving in the space-time, the gravitational field in the entire space-time also change due to this movement. As a consequence, the cosmic time interval, which corresponds to the history of movement of the moving object, changes. This modified the calculation of the real distance S_{real} passed by such a moving object as

$$S_{real}(t) = \int_0^t v(\tau)dg_t(\tau) = {}_0I_t^\alpha v(t), \tag{1.91}$$

where $v(\tau)$ represents the instantaneous velocity of moving object.

Therefore, the fractional order integral of the function $v(t)$ can be interpreted as the real distance traveled by a moving object, for which one can record instantaneous value of its velocity $v(\tau)$ and the instantaneous values of its time τ. Here the relation *between the instantaneous values of time τ (which is homogeneous) and the cosmic time (which is inhomogeneous) is related by the given function $g_t(\tau)$.*

Further, physical interpretation of the Riemann-Lioville fractional derivative is given by Podluby [19] based on the following relation

$$v(t) = {}_0D_t^{\alpha} S_{real}(t), \tag{1.92}$$

and the real velocity $v_{real}(t) = \frac{d}{dt} S_{real}$ based on an independent observer as

$$v_{real}(t) = \frac{d}{dt} {}_0I_t^{\alpha} v(t) = {}_0D_t^{1-\alpha} v(t). \tag{1.93}$$

Therefore, the velocity $v_{real}(t)$ from the viewpoint of the independent observer is the $(1-\alpha)$th *derivative of the instantaneous velocity $v(t)$, if the individual time τ and the cosmic time T are related by $T = g_t(\tau)$. However, for $\alpha = 1$, both velocities are equal because there are no deformations of the time scale.*

1.7.3 Fractional Operators and Linear Filters

In the other way one can interpret Riemann-Lioville's and Weyl's fractional order integration of function $f(t)$ using convolution integrals in which $f(t)$ is convolved with the impulse response of the αth order fractional integrator $h_{-\alpha}(t)$. From the definitions one can observe that for the Riemann integral,

$${}_0I_t^{\alpha} f(t) = \frac{1}{\Gamma(\alpha)} \int_0^t (t-\tau)^{\alpha-1} f(\tau) d\tau, \tag{1.94}$$

and Lioville's definition

$${}_{-\infty}I_t^{\alpha} f(t) = \frac{1}{\Gamma(\alpha)} \int_{-\infty}^t (t-\tau)^{\alpha-1} f(\tau) d\tau, \tag{1.95}$$

the impulse response function is given as

$$h_{-\alpha}(t) = \frac{t^{\alpha-1}}{\Gamma(\alpha)}, \tag{1.96}$$

which is causal. For the Weyl's definition

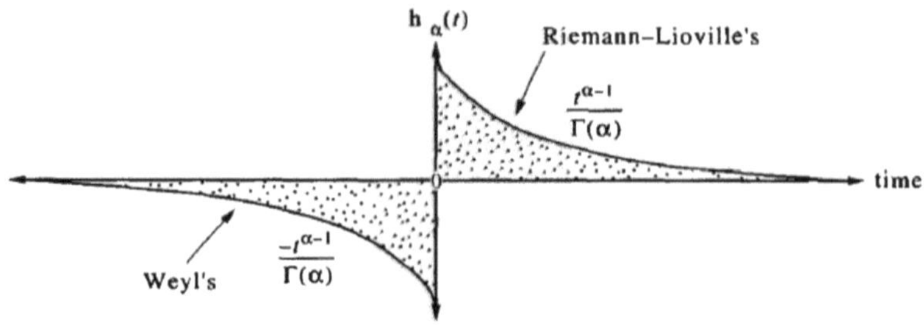

Fig. 1.22 Impulse response function of an αth-order fractional integral

$$
{}_tW_{\infty}^{-\alpha} f(t) = \frac{1}{\Gamma(\alpha)} \int_t^{\infty} (\tau - t)^{\alpha-1} f(\tau) d\tau, \tag{1.97}
$$

the impulse response function is

$$
h_{-\alpha}(t) = \frac{(-t)^{\alpha-1}}{\Gamma(\alpha)}. \tag{1.98}
$$

which is non causal.

In contrast to exponential-type impulse response, the impulse response of the fractional integral is a self similar function as shown in Fig. 1.22, given as

$$
h_{-\alpha}(kt) = \frac{(kt)^{\alpha-1}}{\Gamma(\alpha)} = k^{\alpha-1} h_{-\alpha}(t). \tag{1.99}
$$

Self-similarity means that it is invariant under linear scale change in time. Physical interpretation of fractional order integral is based on the self-similarity property of the convolution integrator in the following way [18]

- For $\alpha = 1$, the impulse response function (1.96), simply becomes a step function. This corresponds to a filter whose impulse response function has a perfect memory which captures the whole history of the input function.
- When $\alpha \to 0$, the impulse response function approaches a one sided delta function. In this way the effect of history is mitigated by the gamma function.
- For the case of $0 < \alpha < 1$, the impulse response function (1.96) decays in a self-similar manner. During the convolution operation of $f(t)$ and $h_{-\alpha}(t)$, past history of function $f(t)$ is partially captured. At each increment of the convolution integral, the recent information of the function is given a maximum weight and the past states are given weights that decrease at a rate proportional to the order of the fractional integration.

For defining βth-order fractional derivative, following relation is used

$$
\begin{aligned}
{}_aD_t^{\beta} f(t) &= \frac{f(0)t^{-\beta}}{\Gamma(1-\beta)} + {}_aD_t^{\beta-1}\dot{f}(t), \;\; 0 \le \beta < 1, \\
&= \frac{f(0)t^{\alpha-1}}{\Gamma(\alpha)} + \frac{1}{\Gamma(\alpha)}\int_a^t \dot{f}(\tau)d\tau,
\end{aligned} \tag{1.100}
$$

where $\dot{f}(t)$ represents the first derivative of the input function and also $\alpha + \beta = 1$. Equation (1.100), in terms of filter can be expressed as

$$
y(t) = h_{-\alpha}(t)f(0) + h_{-\alpha}(t) * \dot{f}(t), \tag{1.101}
$$

where $*$ denotes convolution. From (1.101), one can relate that the βth-order fractional derivative is same as the impulse response of $(1-\beta)$th order fractional integration. One can also note that fractional operators can be regarded as some kind of ordered classes of hereditary integrals. As a direct analogy, one can conclude that a fractional derivative accumulates the history of the lost memory in an evolutionary system with loss.

1.8 First Application of Fractional Calculus or Operators for Physical Problem

The mathematical formulation of fractional calculus was as old as classical calculus but it was not until 1823 that fractional operations were used to solve a specific physical problem. First step in this direction was taken by Niels Henrik Abel [20, 21] who used this new mathematical tool to solve an integral equation arising in the *Tautochrone problem*. In this section we give the detail explanation of *Tautochrone problem* for showing historical and theoretical importance of fractional operator in the real world problem.

Tautochrone problem: A bead of mass m slips along a frictionless wire from height $\eta = y$ at time $t = 0$ to height $\eta = 0$ at time $t = T$. The Tautochrone problem consists of the determination of a smooth curve such that the time required for a particle to slide down the curve to its lowest point under gravity is same for any y.

The tautochrone differential equation is given as

$$
\frac{d\lambda}{\sqrt{y-\eta}} = -\sqrt{2g}dt, \tag{1.102}
$$

where $\lambda = \lambda(\eta)$ is the distance of the particle from the starting point along the curve (or arc-length of curve) and g is the gravitational acceleration.

Proof Using the conservation of energy one can write that the potential energy lost during the descent of the particle is equal to the kinetic energy the particle gains

$$\frac{1}{2}m\left(\frac{d\lambda}{dt}\right)^2 = mg(y-\eta), \tag{1.103}$$

where m represents the mass of particle. Further simplifying,

$$\frac{d\lambda}{dt} = \pm\sqrt{2g(y-\eta)}. \tag{1.104}$$

The positive sign in (1.104) is ignored because the bead speed is in the negative direction of y. Therefore

$$\begin{aligned} \frac{d\lambda}{dt} &= -\sqrt{2g(y-\eta)} \\ \Rightarrow \frac{\lambda'(\eta)d\eta}{\sqrt{y-\eta}} &= -\sqrt{2g}dt. \end{aligned} \tag{1.105}$$

□

The Tautochrone integral equation (curve of the bead's motion) can be obtained by integrating (1.105) as follows

$$\begin{aligned} \int_{\eta=y}^{\eta=0} \frac{\lambda'(\eta)d\eta}{\sqrt{y-\eta}} &= -\sqrt{2g}\int_{t=0}^{t=T} dt \\ \Rightarrow \int_{\eta=0}^{\eta=y} \frac{\lambda'(\eta)d\eta}{\sqrt{y-\eta}} &= \sqrt{2g}T = C. \end{aligned} \tag{1.106}$$

By multiplying both sides of (1.106) with $\Gamma(\frac{1}{2})$, Abel obtained on the right-hand side a fractional integral of order $\frac{1}{2}$

$$D^{-\frac{1}{2}}\lambda'(y) = \frac{C}{\Gamma(\frac{1}{2})}. \tag{1.107}$$

To solve the fractional order Eq. (1.107), differentiate by order $\frac{1}{2}$ on both sides of (1.107)

$$D^{\frac{1}{2}}D^{-\frac{1}{2}}\lambda'(y) = D^{\frac{1}{2}}\frac{C}{\Gamma(\frac{1}{2})}. \tag{1.108}$$

One can also write

$$\lambda'(y) = \frac{C}{\Gamma(\frac{1}{2})} D^{\frac{1}{2}} y^0. \tag{1.109}$$

Lacroix uses the generalized factorial, given by Euler's Gamma function, to formulate non-integer order derivatives as

$$\frac{d^n}{dx^n} x^m = \frac{\Gamma(m+1)}{\Gamma(m-n+1)} x^{m-n}. \tag{1.110}$$

Using (1.110) one can write

$$D^{\frac{1}{2}} y^0 = \frac{\Gamma(0+1)}{\Gamma(0-\frac{1}{2}+1)} y^{-\frac{1}{2}} = \frac{1}{\Gamma(\frac{1}{2})} y^{-\frac{1}{2}}. \tag{1.111}$$

Using (1.109) and (1.111), we get

$$\lambda'(y) = \frac{C}{\Gamma(\frac{1}{2})} \frac{1}{\Gamma(\frac{1}{2})} y^{-\frac{1}{2}} = \frac{C}{\pi \sqrt{y}}. \tag{1.112}$$

Further, one can easily proof that (1.109) represents the equation of *Cycloid.*

Lemma 1.16 *The motion represented as* (1.109) *is along a Cycloid. The Cycloid differential equation is given as*

$$dx = \sqrt{\frac{a}{y} - 1}\, dy, \tag{1.113}$$

where $a = \frac{C^2}{\pi^2}$.

Proof From (1.112), we have

$$\left(\frac{d\lambda}{dy}\right)^2 = \frac{C^2}{\pi^2}\frac{1}{y} = \frac{a}{y}. \tag{1.114}$$

Substituting curve length $d\lambda^2 = (dx)^2 + (dy)^2$, in (1.114)

$$\frac{(dx)^2 + (dy)^2}{(dy)^2} = \frac{a}{y}$$

$$\Rightarrow dx = \sqrt{\frac{a}{y} - 1}\, dy. \tag{1.115}$$

This complete the proof.

Parametric equations of the above motion is given by the following Lemma:

Lemma 1.17 *The parametric equation of Cycloid is given as*

$$x(\theta) = \frac{a}{2}(2\theta + \sin(2\theta))$$
$$y(\theta) = \frac{a}{2}(1 - \cos(2\theta)). \tag{1.116}$$

Proof Integrating (1.115), we obtain

$$x = \int_{\eta=0}^{\eta=y} \sqrt{\frac{a}{n} - 1}\, d\eta. \tag{1.117}$$

Substituting $\eta = a\sin^2\alpha$ and $y = a\sin^2\theta$, we have

$$\sqrt{\frac{a}{n} - 1} = \sqrt{\frac{1}{\sin^2(\alpha)} - 1} = \frac{\cos(\alpha)}{\sin(\alpha)}, \quad d\eta = 2a\sin(\alpha)\cos(\alpha)d\alpha$$

$$\begin{aligned}
\Rightarrow x &= 2a \int_{\alpha=0}^{\alpha=\theta} \cos^2\alpha d\alpha \\
&= a \int_{\alpha=0}^{\alpha=\theta} (1 + \cos(2\alpha))d\alpha \\
&= \frac{a}{2}(2\theta + \sin(2\theta)).
\end{aligned} \tag{1.118}$$

$$y = a\sin^2(\theta) = \frac{a}{2}(1 - cos(2\theta)). \tag{1.119}$$

This complete the proof.

1.8.1 Some More Discussion on Abel Integral and Its Applications

The Able integral can be represented as

$$\frac{1}{\Gamma(\alpha)} \int_0^t \frac{u(\tau)}{(t-\tau)^{1-\alpha}} d\tau = f(t), \tag{1.120}$$

where $0 < \alpha < 1$ and $f(t)$ is a known function. Equation (1.120) can be expressed in terms of fractional integral as

$$I^{\alpha} u(t) = f(t). \tag{1.121}$$

Using the property $D^{\alpha} I^{\alpha} = \mathbb{I}$, where $\mathbb{I}$ represents identity, (1.120) can be written in terms of fractional derivative as

$$u(t) = D^{\alpha} f(t). \tag{1.122}$$

Applying Laplace transform to (1.121)

$$\frac{u(s)}{s^{\alpha}} = f(s)$$
$$\Rightarrow u(s) = s^{\alpha} f(s). \tag{1.123}$$

For applying standard rules of inverse Laplace transform, (1.123) can be expressed in two ways as follows

- Taking s common,

$$u(s) = s\left[\frac{f(s)}{s^{1-\alpha}}\right], \tag{1.124}$$

and applying inverse Laplace transform, we get

$$u(t) = \frac{1}{\Gamma(1-\alpha)} \frac{d}{dt} \int_0^t \frac{f(\tau)}{(t-\tau)^{\alpha}} d\tau. \tag{1.125}$$

- Taking $\frac{1}{s^{1-\alpha}}$ common,

$$u(s) = \frac{1}{s^{1-\alpha}} \left[sf(s) - f(0^+) \right] + \frac{f(0^+)}{s^{1-\alpha}}, \tag{1.126}$$

we obtain

$$u(t) = \frac{1}{\Gamma(1-\alpha)} \frac{d}{dt} \int_0^t \frac{f'(\tau)}{(t-\tau)^{\alpha}} d\tau + f(0^+) \frac{t^{-\alpha}}{\Gamma(1-\alpha)}. \tag{1.127}$$

Thus Niels Henrik Able has given a new platform of fractional order calculus by the solving mechanical problem of the Tautochrone. Problem involved is determining the shape of a curve in the the vertical plane such that the time required for a particle to slide along the curve to its lowest point is equal to a given function of its initial height.

After appropriate changes of variables Able obtained famous integral equation with $\alpha = \frac{1}{2}$, which has been discussed above. Main stress on the problem of isochrone, in which it is required that time of sliding down is independent of the initial height. Able after appropriate transformation of variables noticed that solution of above specified problem is realized only using non-integer order.

It has been found in literature that Able integral equations (more frequently) occurs, in physical situation where measurements are to be evaluated. Some of the most useful applications are listed below

- In evaluation of spectroscopic measurements of cylindrical gas discharges.
- Investigation of star densities in a globular cluster.
- The study of the solar or a planetary atmosphere.
- The inversion of travel times of seismic waves for determination of terrestrial sub-surface structure.
- Spherical stereology.

1.9 Viscoelastic Material and Fractional Calculus

Behavior of viscous liquids under the influence of external forces is given by Newton's law as

$$\sigma(t) = \eta D^1 \varepsilon(t), \tag{1.128}$$

where $\sigma(t)$, $\varepsilon(t)$ and η are stress, strain and viscosity of material respectively. This relation for elastic solids is tackled by Hooke's law as

$$\sigma(t) = E D^0 \varepsilon(t), \tag{1.129}$$

where E is the modulus of elasticity of material. Now consider an experiment where the strain is manipulated in a controlled fashion, say, $\varepsilon(t) = t$ for $t \in [0,\ T]$ with some $T > 0$

$$\begin{aligned} \sigma(t) &= Et, \quad \text{for elastic solid} \\ \sigma(t) &= \eta = Constant, \quad \text{for viscous liquids,} \end{aligned} \tag{1.130}$$

These equations can be summarized as

$$\psi_k = \frac{\sigma(t)}{\varepsilon(t)} t^k, \tag{1.131}$$

where $\psi_0 = E$ and $\psi_1 = \eta$. Evidently the case $k = 0$ corresponds to Hooke's law for solids and $k = 1$ refers to Newton's law for liquids. Viscoelastic material, for example polymers, some types of biological tissue, number of metals aluminium etc., exhibit behavior somewhere between the pure viscous liquid and pure elastic solid. One would observe a relationship of the form (1.130) with $0 < k < 1$. In this case it is appropriate to interprets k as a second material constant in addition to ψ_k. In view of these "interpolation" properties it is natural to assume that it is also possible to model the relation between stress and strain for such a viscoelastic material via an equation of the form

$$\sigma(t) = \upsilon D^k \varepsilon(t), \tag{1.132}$$

where υ is a material constant and $k \in (0, 1)$.

1.10 Application of Fractional Calculus in Econophysics

- Econophysics is a new interdisciplinary area in which the concepts and analysis techniques commonly used for describing physical systems, are applied to investigate the financial and economic problems.
- The dynamics of global markets require full time complete and highly accurate modeling.
- David [22] proposed a very simple model based on the following discussion: **It is well known that a significant change in capital flow commonly occurs when investors at any moment, perceive a minor change in risk.**

Philosophy behind Mathematical Modeling: It seems reasonable to imagine that the square of the capital flow invested, denoted as $(\frac{d\lambda}{dt})^2$, can be proportional to the dimensionless perception of this risk variation, denoted as $(y - y_0)$.
Mathematical Model:

$$\left(\frac{d\lambda}{dt}\right)^2 \propto (y - y_0). \tag{1.133}$$

This can be written as

$$\left(\frac{d\lambda}{dt}\right)^2 = -C(y - y_0). \tag{1.134}$$

An increase in risk perception stimulates a reduction in capital injection. (Negative sign is used in the above equation for this purpose). Also

$$\left(\frac{d\lambda}{dt}\right)^2 = C(y_0 - y). \tag{1.135}$$

This relation can also be considered in the following form

$$\left(\frac{d\lambda}{dt}\right) = C^{\frac{1}{2}}(y_0 - y)^{\frac{1}{2}} \Rightarrow \frac{d\lambda}{(y_0 - y)^{\frac{1}{2}}} = C^{\frac{1}{2}} dt. \tag{1.136}$$

Integrating both sides, one can get

$$\sqrt{C}\int_0^T dt = \int_0^{y_0} (y_0 - y)^{-\frac{1}{2}} d\lambda \Rightarrow k = \int_0^{y_0} (y_0 - y)^{-\frac{1}{2}} d\lambda. \tag{1.137}$$

where $k = \sqrt{C}T$. Here, $\lambda = F(y)$, where λ is the amount of return on capital and y represents the risk perception. Here, $\lambda = F(y)$ and $d\lambda = F'(y)dy$. If we change variable y_0 and y to x and t respectively and replace F' with f, the integral equation becomes

$$k = \int_0^x (x - t)^{-\frac{1}{2}} f(t)dt. \tag{1.138}$$

From this point, the problem is to determine the function $f(t)$. This can be done by multiplying (1.138) by $\frac{1}{\Gamma(\frac{1}{2})}$,

$$\frac{k}{\Gamma(\frac{1}{2})} = \frac{1}{\Gamma(\frac{1}{2})}\int_0^x (x - t)^{-\frac{1}{2}} f(t)dt = {}_0D_x^{-\frac{1}{2}} f(x)$$
$$\Rightarrow {}_0D_x^{\frac{1}{2}} k = \sqrt{\pi} f(x). \tag{1.139}$$

Half order derivative of constant k is written as

$$\begin{aligned} {}_0D_x^{\frac{1}{2}} k &= \frac{d}{dt}\frac{1}{\Gamma(\frac{1}{2})}\int_0^x (x - t)^{-\frac{1}{2}} k dt \\ &= \frac{k}{\Gamma(\frac{1}{2})} x^{-\frac{1}{2}} = \frac{k}{\sqrt{\pi}} x^{-\frac{1}{2}}. \end{aligned} \tag{1.140}$$

Therefore, Eqs. (1.139) and (1.140), it can be concluded that

$$k = \pi\sqrt{x} f(x). \tag{1.141}$$

This model involving risk and capital return based on the non-integer order calculus is simple to implement and offers an interesting alternative for investment and possibly for, predictions in financial markets.

1.11 Age of the Earth Using Heat Equation

The equation which represents the transmission of heat by conduction along single dimensional x is given as

$$\frac{\partial T(x,t)}{\partial t} = d\frac{\partial^2 T(x,t)}{\partial x^2}, \tag{1.142}$$

where d is the diffusivity parameter and $T(x,t)$ is the temperature at point x and at time instant t.

The credit for relating the heat equation for the determination of age of earth goes to Lord Kelvin. The same problem was revisited by Heaviside using the concept of fractional derivative in the following way. Consider uniform initial temperature T_{initial} of the body at time $t = 0$ and assume a constant surface temperature T_{surface}. Considering the variable change $\Psi = T - T_{\text{initial}}$ in (1.142), one can write

$$\frac{\partial \psi(x,t)}{\partial t} = d\frac{\partial^2 \psi(x,t)}{\partial x^2}, \tag{1.143}$$

and the boundary conditions can be reformulated as

$$\psi(x,0) = 0, \tag{1.144}$$

$$\psi(0,t) = T_{\text{surface}} - T_{\text{initial}}. \tag{1.145}$$

Applying Laplace transform to (1.143)

$$\begin{aligned} s\psi(x,s) &= d\frac{\partial^2 \psi(x,s)}{\partial x^2} \\ \Rightarrow \psi &= \lambda_1 e^{x\sqrt{s/d}} + \lambda_2 e^{-x\sqrt{s/d}}, \end{aligned} \tag{1.146}$$

where λ_1 and λ_2 are some constant. Also

$$\frac{\partial \psi(x,s)}{\partial x} = \lambda_1\sqrt{\frac{s}{d}}e^{x\sqrt{s/d}} - \lambda_2\sqrt{\frac{s}{d}}e^{-x\sqrt{s/d}}, \tag{1.147}$$

$$\frac{\partial^2 \psi(x,s)}{\partial x^2} = \lambda_1\frac{s}{d}e^{x\sqrt{s/d}} + \lambda_2\frac{s}{d}e^{-x\sqrt{s/d}}. \tag{1.148}$$

Using (1.147), one can get

$$\frac{\partial \psi(x,s)}{\partial x} = s^{\frac{1}{2}}\frac{1}{\sqrt{d}}\psi(x,s). \tag{1.149}$$

Taking the inverse Laplace transform of (1.149), it can be verified that the temperature gradient at the surface ($x = 0$) satisfies

$$\frac{\partial \psi(0,t)}{\partial x} = \frac{1}{\sqrt{d}}\, {}_0D_t^{1/2} \psi(0,t). \tag{1.150}$$

Heaviside assumed that the radius of the Earth is large enough for the planet to be compared to a semi-infinite solid. It has been also assumed that when Earth was formed, was a molten rock with temperature $\psi(0,0) = 3{,}900\,^{\circ}\mathrm{C}$. One can also further write

$$\frac{\partial \psi(0,0)}{\partial x} = \frac{1}{\sqrt{d}} \frac{3{,}900\,^{\circ}\mathrm{C}}{\sqrt{\pi t}}. \tag{1.151}$$

This expression can be considered to be valid in the future. With the knowledge that $d = 1.178 \times 10^{-6}\,\mathrm{m}^2/\mathrm{s}$ and the temperature gradient at the surface of the earth is 1 °C for every 27.43 m (as measured in the 19th century), the age of earth is $t = 98 \times 10^6$ years.

When we compare the age forecasted by Lord Kelvin and Heaviside, it is found that estimated age given by Kelvin is faulty because he did not take into account heat production due to radioactive decay. However, when we used the fractional order operators to model heat conduction phenomenon, it is found to be accurate due to considering heat emission and conduction.

1.12 Dynamical System and Fractional Order

It is reported in [23] that Chengbin Ma and others have shown that, using a fractional-order model for distributed parameter system is quite natural since the Laplace transform of partial differential equation will inevitably introduce the fractional order s operator. Some of the examples are discussed here.

1.12.1 Cable System

For an electric network containing only the lumped constant, the transit response of the circuit is obtained by the solution of the ordinary integer order differential equation. However, if the system contains the distributed parameters, like communication cables, then the transient response of the circuit cannot be obtained likewise. In this case we have to use partial differential equation and due to this, the differential equation containing non-integer order comes into picture.

Let us calculate the current $i(t, 0)$ which flows into a semi-infinite cable, when a voltage $v(t, 0)$ is applied at one end, as depicted in Fig. 1.23. The equations are written as

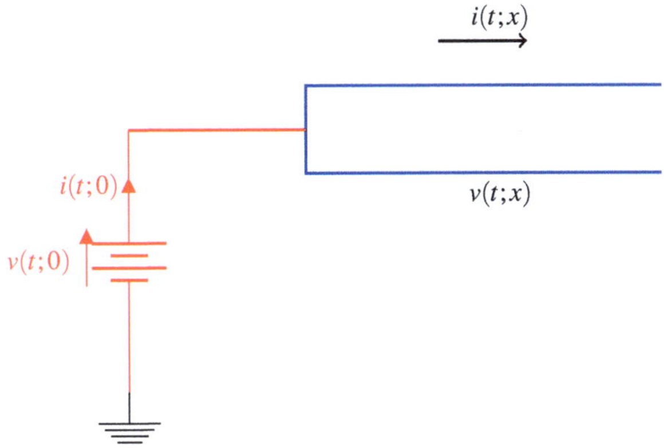

Fig. 1.23 Cable system

$$-\frac{\partial v}{\partial x} = Ri$$
$$-\frac{\partial i}{\partial x} = C\frac{\partial v}{\partial t}, \tag{1.152}$$

where R and C are the resistance and capacitance per unit length respectively. Now, taking the Laplace transform of (1.152), we have

$$\frac{d^2V}{dx^2} = RCsV. \tag{1.153}$$

After solving (1.153) and using the condition that the cable is semi infinite, we have

$$I(s,0) = \sqrt{\frac{C}{R}}\sqrt{s}V(s,0), \tag{1.154}$$

where $I(s,0) = \mathscr{L}\{i(t,0)\}$ and $V(s,0) = \mathscr{L}\{v(t,0)\}$. It is clear from (1.154), that the current is the derivative of the $1/2$-order of the applied voltage.

1.12.2 Flexible Beam

The partial differential equation is used to capture flexible beam dynamics is given as

$$EI\frac{\partial^4 z(x,t)}{\partial x^4} + \rho A\frac{\partial^2 z(x,t)}{\partial t^2} = f(x,t), \tag{1.155}$$

where,

$E :=$ elasticity modulus,
$I :=$ moment of inertia,
$\rho :=$ material density,
$A :=$ beam cross section,
$f(x, t) :=$ external forcing function.

Laplace transform of (1.155) is given as

$$EI\frac{d^4 z(x,s)}{dx^4} + \rho A s^2 z(x,s) = f(x,s). \tag{1.156}$$

Homogeneous solution of (1.156) is given using the following characteristic equation

$$\lambda^4 + \theta s^2 = 0, \tag{1.157}$$

where $\theta = \dfrac{\rho A}{EI}$ and the roots of (1.157) are given as

$$\begin{aligned}
\lambda_1 &= i^{\frac{1}{2}}\theta^{\frac{1}{4}}s^{\frac{1}{2}} \\
\lambda_2 &= -i^{\frac{1}{2}}\theta^{\frac{1}{4}}s^{\frac{1}{2}} \\
\lambda_3 &= i^{\frac{3}{2}}\theta^{\frac{1}{4}}s^{\frac{1}{2}} \\
\lambda_4 &= -i^{\frac{3}{2}}\theta^{\frac{1}{4}}s^{\frac{1}{2}}.
\end{aligned} \tag{1.158}$$

The homogeneous solution which describes the transient response of beam is

$$z(x,s) = \theta_1 e^{\lambda_1 s} + \theta_2 e^{\lambda_2 s} + \theta_3 e^{\lambda_3 s} + \theta_4 e^{\lambda_4 s}, \tag{1.159}$$

where $\theta_1, \theta_2, \theta_3$ and θ_4 are the constants that depend on beam's boundary conditions. From the transient response (1.159), it can be said that it is governed by the fractional order differentiator $s^{1+\frac{1}{2}} = s^{\frac{3}{2}}$, which conforms that flexible beam can be described more precisely using a fractional order transfer function.

1.13 Motion of an Immersed Plate

R.L.Bagley and P.J.Torvik [24] have showed that resulting shear stress at any point in Newtonian fluid can be expressed directly in term of fractional-order derivative of the fluid velocity. This result is depicted in Fig. 1.24. The mathematical exercise carried out by R.L.Bagley and P.J.Torvik is summarized a follows.

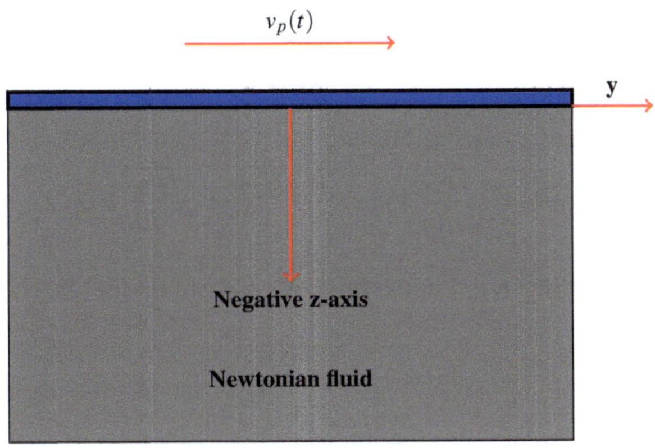

Fig. 1.24 A rigid plate in a Newtonian fluid

The equation of motion of the fluid is

$$\rho_0 \frac{\partial v}{\partial t} = \mu_0 \frac{\partial^2 v}{\partial z^2}, \quad 0 < t < \infty, \quad -\infty < z < 0, \tag{1.160}$$

where ρ_0, μ_0 and $v(t, z)$ are the fluid density, viscosity and transverse velocity respectively. The transverse velocity is a function of time t and the distance z from the fluid plate contact boundary. Let us assume that initially fluid is in equilibrium, i.e.

$$v(0, z) = 0, \quad -\infty < z < 0. \tag{1.161}$$

Also assume that the influence of plate's motion vanishes for $z \to \infty$, i.e.

$$v(t, -\infty) = 0, \;\; 0 < t < \infty. \tag{1.162}$$

Let us consider that the fluid's velocity at $z = 0$ is equal to the given velocity of the plate

$$v(t, 0) = v_p(t). \tag{1.163}$$

Applying Laplace transform, one can obtain

$$\begin{aligned} \rho_0 s V(s, z) &= \mu_0 \frac{d^2 V(s, z)}{dz^2}, \\ V(s, 0) &= V_p(s), \\ V(s, -\infty) &= 0, \end{aligned} \tag{1.164}$$

where $V(s, z) = \mathscr{L}\{v(t, z)\}$ and $V(p) = \mathscr{L}\left\{v_p(t)\right\}$. The solution of (1.164) is

$$V(s,z) = V_p(s)e^{\left(z\sqrt{\frac{\rho_0 s}{\mu_0}}\right)}. \tag{1.165}$$

Differentiating (1.165), we obtain

$$\frac{dV(s,z)}{dz} = \sqrt{\frac{\rho_0 s}{\mu_0}}V_p(s)e^{\left(z\sqrt{\frac{\rho_0 s}{\mu_0}}\right)} = \sqrt{\frac{\rho_0 s}{\mu_0}}V(s,z). \tag{1.166}$$

Using the velocity profile $v(t, z)$ in the fluid, we can obtain the shear stress $\sigma(t, z)$ as

$$\sigma(t,z) = \mu_0 \frac{\partial v(t,z)}{\partial z}. \tag{1.167}$$

Taking the Laplace transform of (1.167) and using (1.166), one can write

$$\sigma^*(s,z) = \mu_0 \frac{dv(s,z)}{dz} = \sqrt{\mu_0 \rho_0 s}\,V(s,z), \tag{1.168}$$

where $\sigma^*(s,z) = \mathscr{L}\{\sigma(t,z)\}$. Now, taking the inverse Laplace transform of (1.168)

$$\sigma(t,z) = \sqrt{\mu_0 \rho_{00}}\,D_t^{1/2} v(t,z). \tag{1.169}$$

Therefore, one can conclude that real physical behavior of system can be expressed in terms of fractional derivative, which was formulated in a conventional manner. The more concrete interpretation of relation (1.169) is that the stress at a given point at any time, in the Newtonian fluid is dependent on the time history of the velocity profile at that point.

Based on the above observation Bagley and Torvik [24] and [25] formulated the following problem :

Statement of Problem: Consider a thin rigid plate of mass m and area S immersed in a Newtonian fluid of infinite extension with density ρ_0 and viscoelastic constant μ_0. The rigid plate is also connected by a massless spring of stiffness k_0 to a fixed point as shown in Fig. 1.25. In addition let us consider that the surface of the plate is sufficiently large so that fluid slows the movement of the plate. Also assume that plate-fluid system initially is in an equilibrium position with zero initial displacement and velocity.

Displacement $x(t)$ of the plate is described by the following differential equation

$$m\ddot{x}(t) = f(t) - k_0 x(t) - 2S\sigma(t,0). \tag{1.170}$$

Now substituting σ from (1.169) in (1.170) and using the relation $v_p(t,0) = \dot{x}(t)$, one can write

$$p\ddot{x}(t) + q\,{}_0D_t^{3/2}x(t) + rx(t) = f(t), \quad t > 0, \tag{1.171}$$

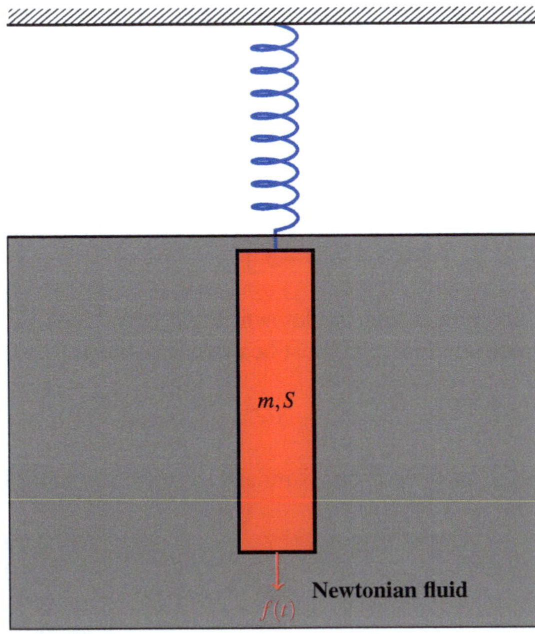

Fig. 1.25 An immersed plate in a Newtonian fluid

where $p = m$, $q = 2S\sqrt{\mu_0\rho_0}$, $r = k_0$, $x(0) = 0$ and $\dot{x}(0) = 0$. Equation (1.171) can be simplified as

$$\ddot{x}(t) + \frac{q}{p}\,{}_0D_t^{3/2}x(t) + \frac{r}{p}x(t) = \frac{f(t)}{p}, \quad t > 0. \tag{1.172}$$

Now consider the following special case of (1.172) [25]

$$\begin{aligned} \ddot{x}(t) + 3{}_0D_t^{3/2}x(t) + x(t) &= f(t), \quad t > 0 \\ x(0) = 0; \quad \dot{x}(0) &= 0, \end{aligned} \tag{1.173}$$

and

$$f(t) = \begin{cases} 8 & \text{if } 0 \le t \le 1 \\ 0 & \text{if } t \ge 1. \end{cases} \tag{1.174}$$

Equation (1.173) can be expressed in the term of sequential linear fractional differential equation as

$${}_0D_t^{4\alpha}x(t) + 3{}_0D_t^{3\alpha}x(t) + x(t) = f(t), \quad \alpha = \frac{1}{2}. \tag{1.175}$$

This equation can be further expressed in the following vectorial form as

$$ {}_0D_t^{\alpha} Y(t) = AY(t) + B, \tag{1.176}$$

where the coordinate have been changed as ${}_0D_t^{i\alpha} x(t) = y_{i+1}$ for $i = 0, 1, 2, 3$. Here

$$ A = \begin{bmatrix} 0 & 1 & 0 & 0 \\ 0 & 0 & 1 & 0 \\ 0 & 0 & 0 & 1 \\ -1 & 0 & 0 & -3 \end{bmatrix}, \quad B = \begin{bmatrix} 0 \\ 0 \\ 0 \\ f(t) \end{bmatrix}, \quad \text{and } Y = \begin{bmatrix} y_1 \\ y_2 \\ y_3 \\ y_4 \end{bmatrix}. \tag{1.177}$$

In addition, using the known initial conditions $x(0) = \dot{x}(0) = 0$ and having the knowledge that $x(t)$ must be differentiable at $[0, t]$, it can be guaranteed that

$$ x(0) = {}_0D_t^{\alpha} y(0) = {}_0D_t^{2\alpha} y(0) = {}_0D_t^{3\alpha} y(0) = 0. \tag{1.178}$$

Let us consider P be a similarity matrix for the digitalization of A, i.e.

$$ \hat{A} = P^{-1}AP = \begin{bmatrix} \lambda_1 & 0 & 0 & 0 \\ 0 & \lambda_2 & 0 & 0 \\ 0 & 0 & \lambda_3 & 0 \\ 0 & 0 & 0 & \lambda_4 \end{bmatrix}, \tag{1.179}$$

where λ_i, $(i = 1, 2, 3, 4)$ are the eigenvalues of A. It is easy to show that

$$ P = \begin{bmatrix} -3.363 + 0.556i & -3.363 - 0.556i & -0.038 & -2.235 \\ -0.911 + 2.073i & -0.911 - 2.073i & 0.114 & 1.709 \\ 0.822 + 1.260i & 0.822 - 1.260i & -0.338 & -1.307 \\ 1 & 1 & 1 & 1 \end{bmatrix}, \tag{1.180}$$

and the eigenvalues of A are

$$ \begin{aligned} \lambda_1 &= 0.363 - 0.556i \\ \lambda_2 &= 0.363 + 0.556i \\ \lambda_3 &= -2.962 \\ \lambda_4 &= -0.765. \end{aligned} \tag{1.181}$$

If we define the transformation as $Y(t) = PZ(t)$ in (1.176), we get

$$ {}_0D_t^{\alpha} Z(t) = \hat{A}Z(t) + \hat{B}, \tag{1.182}$$

where $\hat{B} = P^{-1}B = \hat{b} f(t)$, with

$$\hat{b}f(t) = b_i = \begin{bmatrix} 0.044 - 0.044i \\ 0.044 + 0.044i \\ 1.040 \\ -0.128 \end{bmatrix}, \tag{1.183}$$

with initial condition $Z(0) = 0$. The unique solution of (1.182) is given as

$$Z(t) = 8\int_0^t e_\alpha^{\hat{A}(t-\tau)}\hat{b}d\tau = 8\begin{bmatrix} b_1 \int_0^t e_\alpha^{\lambda_1(t-\tau)}d\tau \\ b_2 \int_0^t e_\alpha^{\lambda_2(t-\tau)}d\tau \\ b_3 \int_0^t e_\alpha^{\lambda_3(t-\tau)}d\tau \\ b_4 \int_0^t e_\alpha^{\lambda_4(t-\tau)}d\tau \end{bmatrix}, \tag{1.184}$$

where

$$e_\alpha^{\hat{A}(t-\tau)} = (t-\tau)^{\alpha-1}\sum_{k=0}^{\infty}\hat{A}^k\frac{(t-\tau)^{k\alpha}}{\Gamma((k+1)\alpha)}, \tag{1.185}$$

and therefore

$$Y(t) = 8P\int_0^t e_\alpha^{\hat{A}(t-\tau)}\hat{b}d\tau = 8P\hat{A}^{-1}\begin{bmatrix} b_1E_\alpha(\lambda_1 t^\alpha) \\ b_2E_\alpha(\lambda_2 t^\alpha) \\ b_3E_\alpha(\lambda_3 t^\alpha) \\ b_4E_\alpha(\lambda_4 t^\alpha) \end{bmatrix}, \tag{1.186}$$

where $E_\alpha(t)$ represents the Mittag-Leffler function, and therefore

$$\begin{aligned} x(t) &= \mathrm{Re}[y_1(t)] \\ &= 10^{-2}\mathrm{Re}\left[(-4.8 + 1.4i)E_\alpha(\lambda_1 t^\alpha) + (0.9 - 7.4i)E_\alpha(\lambda_2 t^\alpha)\right] \\ &\quad + 10^{-2}\mathrm{Re}\left[0.348E_\alpha(\lambda_3 t^\alpha) - 0.59E_\alpha(\lambda_4 t^\alpha)\right]. \end{aligned} \tag{1.187}$$

1.14 Data Fitting and Fractional Order Modeling

Obtaining mathematical models from the measured input and output data is one of the classical problems of literature. The most common practice in many applied field to choose the type of the fitting curve and identify its parameters using some specific criterion. Choosing a particular type of curve means that, the process is modeled by a differential equation, for which the curve is a solution. Some of the examples of data fitting curves are given as follow

- Most simple curve used for data fitting is linear one, which is given as $x(t) = at+b$. If this is the case, then process is modeled by the solution of a simple second order differential equation under two initial conditions, which is expressed as

$$\ddot{x}(t) = 0, \;\; x(0) = b, \;\; \dot{x}(0) = a. \tag{1.188}$$

- Second most common curve is $x(t) = a\sin(\omega t) + b\cos(\omega t)$. If this is the case, then process is modeled by the solution of following second order differential equation under the specified initial conditions

$$\ddot{x}(t) + \omega^2 x(t) = 0, \;\; x(0) = b, \;\; \dot{x}(0) = a\omega. \tag{1.189}$$

- Similarly another frequently used curve is $x(t) = ae^{bt}$. In this case the process is modeled by the solution of following differential equation under the specified initial condition

$$\dot{x}(t) - bx(t) = 0, \;\; x(0) = a. \tag{1.190}$$

All above examples reflect back the behavior of dynamical system by the integer order differential equations. However, most of the time these type of curves are not appropriate for the fitting of input output data. In that particular case data fitting is done using the following more generalized curve [26]

$$x(t) = x(0)E_{\alpha,1}(at^{\alpha}), \tag{1.191}$$

where $E_{\alpha,\beta}(x)$ represents the *Mittag-Leffler* function defined in the following way

$$E_{\alpha,\beta}(x) = \sum_{j=0}^{\infty} \frac{x^j}{\alpha j + \beta}. \tag{1.192}$$

Now our aim is to identify the parameters α, a and x_0. If the data fitting is done using curve (1.191), then the process is modeled by the solution of following fractional order differential equation under the given initial condition

$$ {}_0^C D_t^{\alpha} x(t) - ax(t) = 0, \;\; x(0) = x_0. \tag{1.193}$$

The above procedure gives one of the possible ways to capture fractionality within the systems using input and output measurements and then applying the specified controller, such that the system response will follow some specific behavior.

1.15 Some More Discussion on Properties of Fractional Order Operators

1.15.1 Consistency of Fractional and Integer Order Operators

In the fractional calculus, Caputo derivative and Riemann-Liouville are mostly used. In this subsection, we are going to discuss which fractional order differential operator is consistent with integer order [27]. For the αth order Riemann-Liouville derivative, where $n-1 \le \alpha < n \in \mathbb{Z}^+$

$$ {}_0^{RL}D_t^{\alpha}x(t) = \frac{1}{\Gamma(n-\alpha)}\frac{d^n}{dt^n}\int_0^t (t-\tau)^{n-\alpha-1}x(\tau)d\tau. $$

Now consider the following cases (using the relations $\Gamma(1) = 1, \Gamma(0) = \infty$ and $\Gamma(-k) = \infty$ for $k \in Z^+$)

- when $\alpha \to (n-1)^+$

$$ \begin{aligned} \lim_{\alpha\to(n-1)^+} {}_0^{RL}D_t^{\alpha}x(t) &= \lim_{\alpha\to(n-1)^+}\left(\frac{1}{\Gamma(n-\alpha)}\frac{d^n}{dt^n}\int_0^t (t-\tau)^{n-\alpha-1}x(\tau)d\tau\right) \\ &= \lim_{\alpha\to(n-1)^+}\left(\sum_{k=0}^{n-1}\frac{x^{(k)}(0)t^{-\alpha+k}}{\Gamma(-\alpha+k+1)} + \frac{1}{\Gamma(n-\alpha)}\frac{d^n}{dt^n}\right. \\ &\qquad \left. \times \int_0^t (t-\tau)^{n-\alpha-1}x^{(n)}(\tau)d\tau\right) \\ &= x^{(n-1)}(0) + \int_0^t x^{(n)}(\tau)d\tau = \frac{d^{n-1}x(t)}{dt^{n-1}}, \end{aligned} \tag{1.194} $$

- when $\alpha \to n^-$

$$ \begin{aligned} \lim_{\alpha\to(n)^-} {}_0^{RL}D_t^{\alpha}x(t) &= \lim_{\alpha\to n^-}\left(\frac{1}{\Gamma(n-\alpha)}\frac{d^n}{dt^n}\int_0^t (t-\tau)^{n-\alpha-1}x(\tau)d\tau\right) \\ &= \lim_{\alpha\to n^-}\left(\sum_{k=0}^{n-1}\frac{x^{(k)}(0)t^{-\alpha+k}}{\Gamma(-\alpha+k+1)} + \frac{1}{\Gamma(n-\alpha)}\frac{d^n}{dt^n}\right. \end{aligned} $$

$$
\begin{aligned}
&\times \int_0^t (t-\tau)^{n-\alpha-1} x^{(n)}(\tau)d\tau \Bigg) \\
&= \lim_{\alpha\to n^-}\left(\sum_{k=0}^{n-1}\frac{x^{(k)}(0)t^{-\alpha+k}}{\Gamma(-\alpha+k+1)}+\frac{x^{(n)}(0)t^{n-\alpha}}{\Gamma(n-\alpha+k+1)}\right) \\
&+ \lim_{\alpha\to n^-}\left(\frac{1}{\Gamma(n-\alpha+1)}\frac{d^n}{dt^n}\int_0^t (t-\tau)^{n-\alpha-1}x^{(n)}(\tau)d\tau\right) \\
&= x^{(n)}(0)+\int_0^t x^{(n+1)}(\tau)d\tau = \frac{d^n x(t)}{dt^n}.
\end{aligned}
\tag{1.195}
$$

For the αth order Caputo derivative, where $n-1 \le \alpha < n \in \mathbb{Z}^+$

$$
{}_0^C D_t^\alpha x(t) = \frac{1}{\Gamma(n-\alpha)}\int_0^t (t-\tau)^{n-\alpha-1}x^{(n)}(\tau)d\tau.
$$

Now consider the following cases

- when $\alpha \to (n-1)^+$

$$
\begin{aligned}
\lim_{\alpha\to(n-1)^+} {}_0^C D_t^\alpha x(t) &= \lim_{\alpha\to(n-1)^+}\left(\frac{1}{\Gamma(n-\alpha)}\int_0^t (t-\tau)^{n-\alpha-1}x^{(n)}(\tau)d\tau\right) \\
&= \int_0^t x^{(n)}(\tau)d\tau = x^{(n-1)}(t) - x^{(n-1)}(0),
\end{aligned}
\tag{1.196}
$$

- when $\alpha \to n^-$

$$
\begin{aligned}
\lim_{\alpha\to(n)^-} {}_0^C D_t^\alpha x(t) &= \lim_{\alpha\to n^-}\left(\frac{1}{\Gamma(n-\alpha)}\frac{d^n}{dt^n}\int_0^t (t-\tau)^{n-\alpha-1}x(\tau)d\tau\right) \\
&= \lim_{\alpha\to n^-}\left(\frac{x^{(n)}(0)t^{n-\alpha}}{\Gamma(n-\alpha+1)}+\frac{1}{\Gamma(n-\alpha+1)}\int_0^t (t-\tau)^{n-\alpha}x^{(n+1)}(\tau)d\tau\right) \\
&= x^{(n)}(0)+\int_0^t x^{(n+1)}(\tau)d\tau = \frac{d^n x(t)}{dt^n},
\end{aligned}
\tag{1.197}
$$

Remark 1.18 It is clear from the above mathematical derivation that Riemann-Liouville derivative is a bridge between d^{n-1}/dt^{n-1} and d^n/dt^n. So the

Riemann-Liouville derivative is a reasonable generalization of the classical derivative. However, this is the not the same in the case of Caputo derivative, because Caputo derivative have upper consistency with integer order but not the lower consistency, for $n-1<\alpha<n\in\mathbb{Z}^+$.

1.15.2 Initial Value Problems and Preference of Caputo's Derivative

To make sure the uniqueness of the solutions of classical/fractional differential equations, one needs to specify the initial value conditions. In most of the cases these initial conditions describe certain properties of the solution at the beginning of the process. The initial value problem for the integer and fractional order system is expressed in the following way [27]

- integer order differential equation

$$\begin{cases} x^{(n)}(t)=f(x(t),t), & \text{if } n\in\mathbb{I}^+ \\ x^{(k)}(0)=x_0^k & \text{if } k=0,1,\ldots,n-1. \end{cases} \tag{1.198}$$

- fractional order differential equation using Caputo derivative

$$\begin{cases} {}_0^C D_t^\alpha x(t)=f(x(t),t), & \text{if } n-1<\alpha\leq n\in\mathbb{Z}^+,\ t>0 \\ x^{(k)}(0)=x_0^k, & \text{if } k=0,1,\ldots,n-1. \end{cases} \tag{1.199}$$

- fractional order differential equation using Riemann-Liouville derivative

$$\begin{cases} {}_0^{RL} D_t^\alpha x(t)=f(x(t),t), & \text{if } n-1<\alpha\leq n\in\mathbb{Z}^+,\ t>0 \\ \left[{}_0^{RL} D_t^{\alpha-k} x(t)\right]_{t=0}=\overline{x}_0^k, & \text{if } k=0,1,\ldots,n. \end{cases} \tag{1.200}$$

The initial value problem expressed in (1.198) and (1.199), needs to determine the initial values $x(0),\dot{x}(0),\ldots,x^{(n-1)}$. These initial values in application have clear physical meaning. For example, if one can interprets $x(t)$ as a position , then $\dot{x}(t)$ stands for speed and $\ddot{x}(t)$ expresses as an acceleration. From the mathematical point of view initial value problem expressed in (1.200) is rigorous and elegant. However, it is difficult to give physical meaning of fractional order initial condition. So caputo derivative is more famous among the people who are working in the applied area of fractional calculus.

1.15.3 Why Riemann-Liouville Derivative is More Suitable in Applications?

Actually, Caputo's derivative also has major problems in applications. For example, as shown above, one has

$$\lim_{\substack{\varepsilon>0\\ \varepsilon\to 0}} {}_0^C D_t^{n-\varepsilon} x(t) = x^{(n)}(t)$$
$$\lim_{\substack{\varepsilon>0\\ \varepsilon\to 0}} {}_0^C D_t^{n+\varepsilon} x(t) = x^{(n)}(t) - x^{(n)}(0). \tag{1.201}$$

It implies that due to some system parameters variation, there may be a little change in the n, which may lead to different result in Caputo derivative. If $x^{(n)}(0) \neq 0$, a common case of fractional dynamical system when the system starts from a non still state. Such a problem does not occur in the theory of Riemann-Liouville's derivative.

In addition, when Caputo's fractional-order derivative is applied to describe the constitutive equations of viscoelastic materials, say Nutting's Law as follows

$$\sigma(t) = \nu_0 D_t^{\alpha} \varepsilon(t). \tag{1.202}$$

Then in Caputo derivative a constant strain ε implies that $\sigma = 0$, namely the stress is independent of time t. This claim does not reflect the physical properties of real viscoelastic materials. Where as for Riemann-liouville's derivative, a constant strain ε does not lead to neither zero nor a constant stress which can be easily found if (1.202) is replaced by $\sigma(t) = \nu\, {}_0^{RL} D_t^{\alpha} \varepsilon(t)$. Due to these facts, Riemann-liouville's derivative is probably more suitable for real applications.

1.16 Application of Fractional Calculus in Control

There are lots of applications of fractional calculus in control [28, 29]. It is impossible to talk about every applications. However, some of the applications and their relevant references are given here.

- The control of viscoelastic damped structures [30].
- Control application to a flexible transmission [31].
- Control application to an active suspension [32].
- Control application to a buck converter [33].
- Control application to a hydraulic actuator [34].
- Fractional-order control of a flexible manipulator [35].
- Fractional-order control of a rigid robots [36].
- Fractional-order control of a thermal system [37–39].

1.17 Summary

In this Chapter a brief review of the evolutions of fractional order operators are surveyed. Practical realization and physical interpretations of these operators are also presented. Necessity of fractional order operators for modeling and control have been also discussed. Several practical examples are presented which is better modeled using fractional order operators. Finally discussions of most useful fractional order operators and their limitations have been presented.

References

1. Mandelbrot, B.B.: The Fractal Geometry of Nature. Freeman and Co., San Francisco (1982)
2. Anatoly, A.K., Srivastava, H.M., Trujillo, J.J.: Theory and Applications of Fractional Differential Equations, vol. 204, Elsevier, New York (2006)
3. Ross, B.: Origins of fractional calculus and some applications. Int. J. Math. Statist. Sci **1**(1), 21–34 (1992)
4. Folland, G.B.: Advanced Calculus. Pearson Education India, Delhi (2002)
5. Vinagre, B.M., Chen, Y.Q.: Fractional calculus applications in automatic control and robotics. 41st IEEE CDC2002 Tutorial Workshop
6. Podlubny, I.: Fractional-order systems and pi/sup/spl lambda//d/sup/spl mu//-controllers. IEEE Trans. Autom. Control **44**(1), 208–214 (1999)
7. Vinagre, B.M., Petráš, I., Podlubny, I., Chen, Y.Q.: Using fractional order adjustment rules and fractional order reference models in model-reference adaptive control. Nonlinear Dyn. **29**(1–4), 269–279 (2002)
8. Li, Y., Chen, Y., Podlubny, I.: Mittag-Leffler stability of fractional order nonlinear dynamic systems. Automatica **45**(8), 1965–1969 (2009)
9. Li, C.P., Zhao, Z.G.: Introduction to fractional integrability and differentiability. Eur. Phys. J. Spec. Top. **193**(1), 5–26 (2011)
10. Li, C., Qian, D., Chen, Y.: On Riemann-Liouville and caputo derivatives. Discrete Dyn. Nat. Soc., 562494, pp. 1–15 (2011)
11. Kolwankar, K.M., Gangal, A.D.: Fractional differentiability of nowhere differentiable functions and dimensions. Chaos: an Interdisciplinary. J. Nonlinear Sci. **6**(4), 505–513 (1996)
12. Podlubny, I.: Fractional Differential Equations: An Introduction to Fractional Derivatives, Fractional Differential Equations, to Methods of Their Solution and Some of Their Applications, vol. 198, Access Online via Elsevier (1998)
13. Zhang, X., Chen, Y.Q.: Remarks on fractional order control systems. In: American Control Conference (ACC), IEEE, 2012, pp. 5169–5173. (2012)
14. Oldham, K.B., Spanier, J.: The Fractional Calculus: Theory and Applications of Differentiation and Integration to Arbitrary Order, vol. 111, Academic Press, New York (1974)
15. Oldham, K.B., Zoski, C.G.: Analogue instrumentation for processing polarographic data. J. Electroanal. Chem. Interfacial Electrochem. **157**(1), 27–51 (1983)
16. Wall, H.S.: Analytic Theory of Continued Fractions. New York: Chelsea (1948)
17. Machado, J.A.T.: A probabilistic interpretation of the fractional-order differentiation. Fractional Calc. Appl. Anal. **6**(1), 73–80 (2003)
18. Moshrefi-Torbati, M., Hammond, J.K.: Physical and geometrical interpretation of fractional operators. J. Frankl. Inst. **335**(6), 1077–1086 (1998)
19. Podlubny, I.: Geometric and physical interpretation of fractional integration and fractional differentiation. arXiv preprint math/0110241 (2001)

20. Gorenflo, R.: Abel integral equations with special emphasis on applications. Lectures in Mathematical Sciences 13 (1996)
21. Debnath, L.: A brief historical introduction to fractional calculus. Int. J. Math. Educ. Sci. Technol. **35**(4), 487–501 (2004)
22. David, S., Linares, J., Pallone, E.: Fractional order calculus: historical apologia, basic concepts and some applications. Revista Brasileira de Ensino de Física **33**(4), 4302–4302 (2011)
23. O'Connor, W.J.: Control of flexible mechanical systems: wave-based techniques. In: American Control Conference, 2007. ACC'07, IEEE, pp. 4192–4202 (2007)
24. Torvik, P.J., Bagley, R.L.: On the appearance of the fractional derivative in the behavior of real materials. J. Appl. Mech. **51**(2), 294–298 (1984)
25. Bonilla, B., Rivero, M., Trujillo, J.J.: On systems of linear fractional differential equations with constant coefficients. Appl. Math. Comput. **187**(1), 68–78 (2007)
26. Sierociuk, D., Podlubny, I., Petras, I.: Experimental evidence of variable-order behavior of ladders and nested ladders. IEEE Trans. Control Syst. Technol. **21**(2), 459–466 (2013)
27. Li, C., Deng, W.: Remarks on fractional derivatives. Appl. Math. Comput. **187**(2), 777–784 (2007)
28. Debnath, L.: Recent applications of fractional calculus to science and engineering. Int. J. Math. Math. Sci. **2003**(54), 3413–3442 (2003)
29. Ma, C., Hori, Y.: Fractional-order control: theory and applications in motion control. IEEE Ind. Electron. Mag. **1**(4), 6–16 (2007)
30. Sánchez, Y.: Fractional-pid control for active reduction of vertical tail buffeting. Technical Report (1999)
31. Oustaloup, A., Mathieu, B., Lanusse, P.: The crone control of resonant plants: application to a flexible transmission. Eur. J. Control **1**(2), 113–121 (1995)
32. Lanusse, P., Poinot, T., Cois, O., Oustaloup, A., Trigeassou, J.: Tuning of an active suspension system using a fractional controller and a closed-loop tuning. In: 11th International Conference on Advanced Robotics, pp. 258–263 (2003)
33. Calderón, A., Vinagre, B., Feliu, V.: Linear fractional order control of a dc-dc buck converter. In: ECC 2003: European control conference (2003)
34. Pommier-Budinger, V., Musset, R., Lanusse, P., Oustaloup, A.: Study of two robust controls for an hydraulic actuator. In: European Control Conference, Cambridge, UK, pp.1–4 (2003)
35. Monje, C., Ramos, F., Feliu, V., Vinagre, B.: Tip position control of a lightweight flexible manipulator using a fractional order controller. IET Control Theory Appl. **1**(5), 1451–1460 (2007)
36. Ferreira, N.F., Machado, J.T.: Fractional-order hybrid control of robotic manipulators. In: Proceedings of the 11th International Conference on Advanced Robotics, vol. 398, IEEE Press, Piscataway (2003)
37. Vinagre, B., Petras, I., Merchan, P., Dorcak, L.: Two digital realizations of fractional controllers: Application to temperature control of a solid. In: Proceedings of the European Control Conference (ECC2001), Porto, pp. 1764–1767 (2001)
38. Petráš, I., Vinagre, B.M.: Practical application of digital fractional-order controller to temperature control. Acta Montan Slovaca **7**(2), 131–137 (2002)
39. Petráš, I., Vinagre, B.M., Dorčák, L., Feliu, V.: Fractional digital control of a heat solid–experimental results. In: International Carpathian Control Conference, Malenovice, pp. 365–370 (2002)

Chapter 2
Solution, Stability and Realization of Fractional Order Differential Equation

2.1 Introduction

Classical calculus has provided an efficient tool for modeling and exploring the properties of the dynamical system problems concerning of physics, biology, engineering and applied sciences. However, experiments with a realistic approach teach us that there are a large class of complex systems where microscopic and macroscopic behaviors are not captured or properly explained using classical calculus. Some examples can be stated: relaxation in viscoelastic materials like polymers, the spread of contaminants in underground water, network traffic, charge transport in amorphous semiconductors, cell diffusion process, the transmission of signals through strong magnetic fields such as those found within confined plasma etc. After several years of research and discussion, it has been found that these major classes of complex systems which contains non-local dynamics involving long-memory are captured using a more general class of operators known as fractional operators. The differential equations involving these operators are known as fractional order differential equation.

Stability is the one of the most frequent terms used in literature whenever we deal with the dynamical systems and their behaviors. In mathematical terminology, stability theory addresses the convergence of solutions of differential or difference equations and of trajectories of dynamical systems under small perturbations of initial conditions. Same as classical differential or difference equations a lot of stress has been given to the stability and stabilization of the systems represented by fractional order differential equations.

Up to this point it is quit obvious that the fractional order calculus is more appropriate to capture the real dynamical behavior rather than integer order calculus. However, fractional order systems have an infinite dimension, while the integer order systems is only finite dimensional. Therefore, to realize the fractional order controllers perfectly, all past inputs should be memorized, which is not possible without proper approximation. This issue is also discussed in this chapter.

B. Bandyopadhyay and S. Kamal, *Stabilization and Control of Fractional Order Systems: A Sliding Mode Approach*, Lecture Notes in Electrical Engineering 317,
DOI 10.1007/978-3-319-08621-7_2

The brief outline of this chapter is as follows. Section 2.2 describes the solution of fractional differential equations and Mittag-Leffler function. Section 2.3 discusses the brief summary of the notation of stability and stabilization. A brief review on linear matrix inequality (LMI) stability conditions for LTI fractional order systems are analyzed in Sect. 2.4. A deep discussion on the realization issue of fractional-order controller is presented in Sect. 2.5. A brief review of fractional order PID control is surveyed in Sect. 2.6 followed by the concluding Sect. 2.7.

2.2 Solution of Fractional Differential Equations and Mittag-Leffler Function

As already discussed in Chap. 1 the most popular fractional order derivative has been given by Riemann-Liouville and Caputo. Therefore, in this book we concentrate on fractional order differential equations formed by these two derivatives only. Riemann-Liouville's fractional-order derivative has been most widely used for capturing the physical problems because it places less constraints on the concerned function. However, fractional order differential equations involving Riemann-Liouville's fractional-order derivative has some practical issues, related to initial value problem. This is because the initial problems contain the fractional operator which does not have a straightforward physical meaning. Initial value problem for a non-homogeneous fractional differential equation under non-zero initial conditions, is expressed as [1, 2]

$$ {}_0D_t^{\alpha}x(t) - \lambda x(t) = f(t, x(t)), \quad \left[{}_0D_t^{\alpha-k}x(t)\right]_{t=0} = c_k \quad (k = 1, 2, \ldots, n), $$

where $n - 1 < \alpha < n$. However, in the case of Caputo's fractional differential equation, the initial value problem can be represented as

$$ {}_0D_t^{\alpha}x(t) - \lambda x(t) = f(t, x(t)), \quad \left[{}_0D_t^{k}x(t)\right]_{t=0} = \hat{c}_k \quad (k = 0, 1, 2, \ldots, n-1). $$

When we are solving fractional order differential equations and fractional order integral equations, the *Mittag-Leffler* function comes into picture. Therefore, before presenting the solution of fractional differential equations, we begin with the discussions on definition and various properties of *Mittag-Leffler* functions.

2.2.1 *Mittag-Leffler Function*

The function $E_{\alpha}(t)$ was defined by Mittag-Leffler in the year 1903. It is a direct generalization of the exponential series [1, 2]. For $\alpha = 1$ we have the exponential series.

Definition 2.1 The Mittag-Leffler function function $E_{\alpha}(t)$ and the generalized Mittag-Leffler function $E_{\alpha,\beta}(t)$ are defined as:

$$E_{\alpha}(t) = \sum_{k=0}^{\infty} \frac{t^k}{\Gamma(\alpha k + 1)}, \quad \alpha > 0. \tag{2.1}$$

For $\alpha = 1$, we have the exponential series. Similarly,

$$E_{\alpha,\ \beta}(t) = \sum_{k=0}^{\infty} \frac{t^k}{\Gamma(\alpha k + \beta)}, \quad \alpha, \beta > 0. \tag{2.2}$$

The other well known function is *Miller-Ross function* which is defined as:

Definition 2.2

$$\xi_{\alpha,\ a}(t) = \sum_{k=0}^{\infty} \frac{a^k t^{k+\alpha}}{\Gamma(\alpha + k + 1)} = t^{\alpha} E_{1,\alpha+1}(at). \tag{2.3}$$

Some special cases of Mittag-leffler functions are summarized as,

$$\xi_{0,1}(t) = E_1(t) = E_{1,1}(t) = \sum_{k=0}^{\infty} \frac{t^k}{\Gamma(k+1)} = \sum_{k=0}^{\infty} \frac{t^k}{k!} = e^t, \tag{2.4}$$

$$\xi_{0,\ a}(t) = E_1(at) = E_{1,1}(at) = e^{at}. \tag{2.5}$$

The following Laplace transform is involved frequently while solving fractional differential equation:

Theorem 2.3 *The Laplace transform of* $t^{\alpha k+\beta-1} \frac{d^k E_{\alpha,\beta}(\pm a t^{\alpha})}{d(\pm a t^{\alpha})^k}$,

$$\mathscr{L}\left[t^{\alpha k+\beta-1} \frac{d^k E_{\alpha,\beta}(\pm a t^{\alpha})}{d(\pm a t^{\alpha})^k} \right] = \frac{k! s^{\alpha-\beta}}{(s^{\alpha} \mp a)^{k+1}}. \tag{2.6}$$

Proof Consider the following integral

$$\begin{aligned} \int_0^{\infty} e^{-t} t^{\beta-1} E_{\alpha,\beta}(\pm \eta t^{\alpha}) dt &= \int_0^{\infty} e^{-t} t^{\beta-1} \sum_{k=0}^{\infty} \frac{(\pm\eta)^k t^{\alpha k}}{\Gamma(\alpha k + \beta)} dt \\ &= \sum_{k=0}^{\infty} \frac{(\pm\eta)^k}{\Gamma(\alpha k + \beta)} \int_0^{\infty} e^{-t} t^{\alpha k+\beta-1} dt. \end{aligned} \tag{2.7}$$

Using the following relation

$$\int_0^\infty e^{-t} t^{\alpha k+\beta-1} dt = \Gamma(\alpha k+\beta),$$

(2.7) can be written as,

$$\int_0^\infty e^{-t} t^{\beta-1} E_{\alpha,\beta}(\pm\eta t^\alpha) dt = \sum_{k=0}^{\infty} \frac{(\pm\eta)^k}{\Gamma(\alpha k+\beta)} \Gamma(\alpha k+\beta)$$
$$= \frac{1}{1\mp\eta}. \tag{2.8}$$

kth differentiation of (2.8) is given as

$$\frac{d^k}{d\eta^k} \int_0^\infty e^{-t} t^{\beta-1} E_{\alpha,\beta}(\pm\eta t^\alpha) dt = \frac{k!(\pm 1)^k}{(1\mp\eta)^{k+1}}$$

$$\int_0^\infty e^{-t} t^{\beta-1} (\pm t^\alpha)^k \frac{d^k}{d(\pm\eta t^\alpha)^k} E_{\alpha,\beta}(\pm\eta t^\alpha) dt = \frac{k!(\pm 1)^k}{(1\mp\eta)^{k+1}}. \tag{2.9}$$

Now changing the variable t with st

$$\frac{k!(\pm 1)^k}{(1\mp\eta)^{k+1}} = \int_0^\infty e^{-st} s^{\beta-1} t^{\beta-1} (\pm 1)^k s^{\alpha k} t^{\alpha k} \frac{d^k E_{\alpha,\beta}(\pm\eta s^\alpha t^\alpha)}{d(\pm\eta s^\alpha t^\alpha)^k} s dt. \tag{2.10}$$

Also by replacing ηs^α by a

$$\frac{k!}{s^\beta s^{\alpha k}\left(1\mp\frac{a}{s^\alpha}\right)^{k+1}} = \int_0^\infty e^{-st} t^{\alpha k+\beta-1} \frac{d^k E_{\alpha,\beta}(\pm a t^\alpha)}{d(\pm a t^\alpha)^k} dt. \tag{2.11}$$

This completes the proof.

Some special cases of the Laplace transform (2.6) :

- Substituting $k=0$ in (2.6)

$$\mathscr{L}\left[t^{\beta-1} E_{\alpha,\beta}(\pm a t^\alpha)\right] = \frac{s^{\alpha-\beta}}{s^\alpha \mp a}. \tag{2.12}$$

- Substituting $\alpha = \beta$ in (2.12)

$$\mathscr{L}\left[t^{\alpha-1}E_{\alpha,\alpha}(\pm at^{\alpha})\right] = \frac{1}{s^{\alpha} \mp a}. \tag{2.13}$$

- *Example* Inverse Laplace transform of $\frac{4s^{1/2}-1}{s+s^{1/2}-2}$ using (2.13)

$$\begin{aligned}\mathscr{L}^{-1}\left[\frac{4s^{1/2}-1}{s+s^{1/2}-2}\right] &= \mathscr{L}^{-1}\left[\frac{1}{s^{1/2}-1}+\frac{3}{s^{1/2}+2}\right]\\ &= t^{-1/2}E_{1/2,1/2}(t^{1/2}) + 3t^{-1/2}E_{1/2,1/2}(-2t^{1/2}).\end{aligned} \tag{2.14}$$

2.2.2 Solution the Fractional Differential Using Laplace Transform

Consider the following fractional differential equation,

$$\begin{aligned}{}_0D_t^{\alpha}x(t) - \lambda x(t) = f(t, x(t)),\ &\left[{}_0D_t^{\alpha-k}x(t)\right]_{t=0}\\ &= c_k \quad (k = 1, 2, \ldots, n).\end{aligned} \tag{2.15}$$

Laplace transform of Riemann-Liouville's derivative is given as,

$$\int_0^{\infty} e^{-st}{}_0D_t^{\alpha}x(t)dt = s^{\alpha}X(s) - \sum_{k=0}^{n-1} s^k\left[{}_0D_t^{\alpha-k-1}x(t)\right]_{t=0}.$$

Taking Laplace transform of (2.15), one can write

$$s^{\alpha}X(s) - \lambda X(s) = F(s) + \sum_{k=1}^{n} c_k s^{k-1}$$

$$X(s) = \frac{F(s)}{(s^{\alpha}-\lambda)} + \sum_{k=1}^{n} c_k \frac{s^{k-1}}{(s^{\alpha}-\lambda)}. \tag{2.16}$$

Using (2.6) the inverse Laplace transformation of (2.16) can be found as,

$$x(t) = \sum_{k=1}^{n} c_k t^{\alpha-k}E_{\alpha,\alpha-k+1}(\lambda t^{\alpha}) + \int_0^t (t-\tau)^{\alpha-1}E_{\alpha,\alpha}(\lambda(t-\tau)^{\alpha})f(\tau)d\tau. \tag{2.17}$$

Example Consider an example of fractional differential equation as,

$$ {}_0D_t^{\frac{1}{2}}x(t) + ax(t) = 0, \quad (t > 0); \quad \left[{}_0D_t^{-1/2}x(t)\right]_{t=0} = C. \tag{2.18} $$

Applying Laplace transform, one can write

$$ X(s) = \frac{C}{s^{1/2} + a}. \tag{2.19} $$

Taking inverse Laplace transform, one can write solution of (2.19), as discussed in (2.17) as

$$ x(t) = Ct^{-1/2}E_{1/2,1/2}(-a\sqrt{t}). \tag{2.20} $$

2.2.3 More Proper Way to Impose Initial Condition to Fractional Order Differential Equation

Initialized fractional order Riemann-Liouville derivative is expressed in the following way

$$ \begin{aligned} {}_{t_0}D_t^{\alpha} f(t) &= \frac{1}{\Gamma(n-\alpha)} \frac{d^n}{dt^n} \int_{t_0}^{t} \frac{f(\tau)}{(t-\tau)^{1+\alpha-n}} d\tau, \quad t > a \\ f &= \varphi(t), \quad f \in C^{n-1}, \quad t_0 < t \le a \end{aligned} \tag{2.21} $$

where n is a positive integer satisfying $n - 1 < \alpha \le n$, and $f = \varphi(t)$ represents the initial history over $(t_0, a]$, and $f = 0$, for $t \le t_0$.

For visualizing the effect of initial history Du and Wang [3] considered the example of axially loaded viscoelastic bar, of which elongation $x(t)$ and longitudinal force F satisfy the following equation

$$ F = {}_0D_t^{0.5}x(t). \tag{2.22} $$

Suppose that the elongation $x(t)$ is initialized as $x(t) = \varphi(t) = t$ for $t \in (0, 1]$ and it is kept constant $(x(t) = 1)$ when $t > 1$; then the force can be calculated as

$$ \begin{aligned} F &= {}_0D_1^{0.5}x(t) + {}_1D_t^{0.5}x(t) \\ &= \frac{1}{\Gamma(0.5)} \frac{d}{dt} \left[\int_0^1 \frac{\tau}{(t-\tau)^{0.5}} d\tau + \int_1^t \frac{1}{(t-\tau)^{0.5}} d\tau \right] \\ &= \frac{2(\sqrt{t} - \sqrt{t-1})}{\Gamma(0.5)}. \end{aligned} \tag{2.23} $$

Using (2.23) it can be concluded that viscoelastic force is still dependent on t, although the elongation $x(t)$ is constant after $t > 1$. It can be also seen that the force is sensitive to the initial history. If the initial history is given as $\varphi(t) = \sqrt{t}$, then the force is given as

$$F = \frac{1}{\Gamma(0.5)} \sin^{-1}\left(\frac{1}{\sqrt{t}}\right). \tag{2.24}$$

In both the cases one can see that the net viscoelastic force depends on the initial history.

Very often the initial value problem of a fractional differential equation is converted to an equivalent integral equation. Consider the following fractional order differential equation

$$\begin{aligned} {}_{t_0}D_t^{\alpha} x(t) &= f(t, x(t)), \quad (t > a) \\ x(t) &= \varphi(t_0, a], \ x_0 \in C^n. \end{aligned} \tag{2.25}$$

Theorem 2.4 *The initial value problem* (2.25) *is equivalent to the following integral equation*

$$x(t) = \begin{cases} F(\varphi(t); t, x(t)) & \text{if } t > a \\ \varphi(t) & \text{if } t_0 < t \leq a \end{cases} \tag{2.26}$$

where

$$\begin{aligned} F(\varphi(t); t, x(t)) &= {}_aD_t^{-\alpha} f(t, x(t)) - {}_aD_t^{n-\alpha}\left[{}_{t_0}D_a^{-(n-\alpha)}\varphi(t)\right] \\ &+ \sum_{j=1}^{n} \frac{(t-a)^{\alpha-j}}{\Gamma(\alpha-j+1)} \left[{}_{t_0}D_t^{(\alpha-j)}\varphi(t)\right]_{t=a}. \end{aligned} \tag{2.27}$$

Proof Following Lemma is important before proving the main result which is given by Du and Wang [3].

Lemma 2.5 *For a given $\alpha > 0$, one has*

$$\begin{aligned} {}_aD_t^{-\alpha}\left[{}_{t_0}D_t^{\alpha} x(t)\right] &= x(t) + {}_aD_t^{n-\alpha}\left[{}_{t_0}D_a^{-(n-\alpha)} x(t)\right] \\ &- \sum_{j=1}^{m} \frac{(t-a)^{\alpha-j}}{\Gamma(\alpha-j+1)} \left[{}_{t_0}D_t^{(\alpha-j)} x(t)\right]_{t=a}. \end{aligned} \tag{2.28}$$

Proof Let

$$L(t) = \frac{1}{\Gamma(n-\alpha)} \int_{t_0}^{t} \frac{x(\tau)}{(t-\tau)^{1+\alpha-n}} d\tau = {}_{t_0}D_t^{-(n-\alpha)} x(t), \tag{2.29}$$

then

$$L^{(n)}(t) = {}_{t_0}D_t^{\alpha} x(t), \tag{2.30}$$

$$L^{(n-j)}(a) = \left[L^{(n-j)}(t)\right]_{t=a} = \left[{}_{t_0}D_t^{\alpha-j} \varphi(t)\right]_{t=a}. \tag{2.31}$$

Now

$$\begin{aligned}
{}_aD_t^{-\alpha}\, {}_{t_0}D_t^{\alpha} x(t) &= \frac{d}{dt}\, {}_aD_t^{-\alpha-1}\, {}_{t_0}D_t^{\alpha} x(t) \\
&= \frac{d}{dt}\left[\frac{1}{\Gamma(\alpha+1)} \int_a^t (t-\tau)^{\alpha} L^{(n)}(\tau) d\tau\right] \\
&= \frac{d}{dt}\left[\frac{1}{\Gamma(\alpha+1-n)} \int_a^t \frac{L(\tau)}{(t-\tau)^{n-\alpha}} d\tau - \sum_{j=1}^{n} \frac{(t-a)^{\alpha-j+1}}{\Gamma(\alpha-j+2)} L^{(n-j)}(a)\right] \\
&= \frac{d}{dt}\, {}_aD_t^{-(\alpha-n+1)} L(t) - \sum_{j=1}^{n} \frac{(t-a)^{\alpha-j}}{\Gamma(\alpha-j+1)} L^{(n-j)}(a).
\end{aligned} \tag{2.32}$$

As $L(t) = \left({}_{t_0}D_a^{-(n-\alpha)} + {}_aD_t^{-(n-\alpha)}\right) x(t)$, by substituting $L(t)$ in (2.32) one can write

$$\begin{aligned}
{}_aD_t^{-\alpha}\, {}_{t_0}D_t^{\alpha} x(t) &= \frac{d}{dt}\left[{}_aD_t^{-1} x(t) + {}_aD_t^{-(\alpha-n+1)}\, {}_{t_0}D_{\varphi}^{-(n-\alpha)}(a)\right] \\
&\quad - \sum_{j=1}^{n} \frac{(t-a)^{\alpha-j}}{\Gamma(\alpha-j+1)} L^{(n-j)}(a) \\
&= x(t) + \frac{d}{dt}\left[{}_aD_t^{-(\alpha-n+1)}\, {}_{t_0}D_a^{-(n-\alpha)} \varphi(t)\right] \\
&\quad - \sum_{j=1}^{n} \frac{(t-a)^{\alpha-j}}{\Gamma(\alpha-j+1)} \left[{}_{t_0}D_t^{(\alpha-j)} \varphi(t)\right]_{t=a} \\
&= x(t) + \left[{}_aD_t^{(n-\alpha)}\, {}_{t_0}D_a^{-(n-\alpha)} \varphi(t)\right] - \sum_{j=1}^{n} \frac{(t-a)^{\alpha-j}}{\Gamma(\alpha-j+1)} \left[{}_{t_0}D_t^{(\alpha-j)} \varphi(t)\right]_{t=a}.
\end{aligned} \tag{2.33}$$

This proves the Lemma.

If $x(t)$ is the solution of (2.25), then

$$ {}_aD_t^{-\alpha}\,{}_{t_0}D_t^{\alpha}x(t) = {}_aD_t^{-\alpha}f(t,x(t)). \tag{2.34}$$

It can be verified that $x(t)$ satisfies (2.26), due to above Lemma.

Conversely, if $x(t)$ is a solution of (2.26), then

$$ \begin{aligned} {}_aD_t^{\alpha}x(t) &= f(t,x(t)) - {}_aD_t^{\alpha}\left({}_aD_t^{n-\alpha}\left[{}_{t_0}D_a^{-(n-\alpha)}\varphi(t)\right]\right) \\ &+ {}_aD_t^{\alpha}\left(\sum_{j=1}^{n}\frac{(t-a)^{\alpha-j}}{\Gamma(\alpha-j+1)}\left[{}_{t_0}D_t^{(\alpha-j)}\varphi(t)\right]_{t=a}\right). \end{aligned} \tag{2.35}$$

Adding ${}_{t_0}D_a^{\alpha}x(t)$ to both sides and, since ${}_aD_t^{\alpha}(t-a)^{\alpha-j} = 0$ for $j = 1, 2, \ldots, n$. Equation (2.35) is simplified as,

$$ \begin{aligned} {}_{t_0}D_t^{\alpha}x(t) &= f(t,x(t)) + {}_{t_0}D_a^{\alpha}x(t) - {}_aD_t^{\alpha}\left({}_aD_t^{n-\alpha}\left[{}_{t_0}D_a^{-(n-\alpha)}\varphi(t)\right]\right) \\ &+ {}_aD_t^{\alpha}\left(\sum_{j=1}^{n}\frac{(t-a)^{\alpha-j}}{\Gamma(\alpha-j+1)}\left[{}_{t_0}D_t^{(\alpha-j)}\varphi(t)\right]_{t=a}\right) \\ &= f(t,x(t)) + {}_{t_0}D_a^{\alpha}x(t) - {}_aD_t^{\alpha}\left({}_aD_t^{n-\alpha}\left[{}_{t_0}D_a^{-(n-\alpha)}\varphi(t)\right]\right). \end{aligned} \tag{2.36}$$

Also, one can write

$$ {}_{t_0}D_a^{\alpha}x(t) = \frac{d^n}{dt^n}\,{}_{t_0}D_a^{-(n-\alpha)}\varphi(t) = {}_aD_t^{\alpha}\left({}_aD_t^{n-\alpha}\left[{}_{t_0}D_a^{-(n-\alpha)}\varphi(t)\right]\right). \tag{2.37}$$

Thus, ${}_{t_0}D_t^{\alpha}x(t) = f(t,x(t))$, $t > a$. It implies that $x(t)$ is a solution of (2.25). This completes the proof.

Remark 2.6 The initial value problem of fractional differential equation can be converted to an equivalent integral equation and it is easy for both theoretical and numerical analysis.

2.3 Stability and Stabilization

Control system problems generally cater to two categories, first is the stabilization or regulation and second is the tracking or servo. Stabilization problems, aim to design a control system, known as stabilizer or a regulator, so that the state of the closed-loop system will be stabilized around the desired point also known as an equilibrium point. In tracking problems, the design objective is to construct a suitable controller, called a tracker, so that the system output tracks a given time-varying reference trajectory. When we see the tracking problem in the frame of difference between reference trajectory and trajectories generated by system, which is called error; then tracking problem is converted into a stabilization problem of the error variable. Therefore, in this book we restrict our analysis on stabilization problem.

2.3.1 Concept of Equilibrium Point

Equilibrium point of fractional order system is defined same as in integer order [4]. Consider the Riemann-Liouville fractional order autonomous system

$$ {}_{t_0}^{RL}D_t^{\alpha} x(t) = f(t, x), \tag{2.38}$$

with initial condition $x(t_0)$, where $\alpha \in (0\ \ 1)$, $f : [t_o, \infty] \times \boldsymbol{\Omega} \to \mathbb{R}^n$ is piecewise continuous in t and locally Lipschitz in x on $[t_o, \infty] \times \boldsymbol{\Omega}$, $\boldsymbol{\Omega} \in \mathbb{R}^n$ is a domain that contains the equilibrium point $x = 0$.

Definition 2.7 The constant x_0 is an equilibrium point of the Riemann-Liouville fractional dynamic system (2.38), if and only if

$$ {}_{t_0}^{RL}D_t^{\alpha} x_0 = f(t, x_0). \tag{2.39}$$

Just like integer order, shifting of equilibrium point is valid for fractional order. Without loss of generality any equilibrium point can be shifted to origin via a change of variables. Suppose the equilibrium point for (2.38) is $\bar{x} \neq 0$ and consider the change of variable $y = x - \bar{x}$. The αth order derivative of y is given by Riemann-Liouville fractional order autonomous system

$$\begin{aligned} {}_{t_0}^{RL}D_t^{\alpha} y = {}_{t_0}^{RL}D_t^{\alpha}(x - \bar{x}) &= f(t, x) - \frac{\bar{x}t^{-\alpha}}{\Gamma(1-\alpha)} \\ &= f(t, y + \bar{x}) - \frac{\bar{x}t^{-\alpha}}{\Gamma(1-\alpha)} = \bar{g}(t, y), \end{aligned} \tag{2.40}$$

$\bar{g}(t, 0) = 0$ and in terms of the new variable y, the system has equilibrium at the origin.

Another popular definition used to represent dynamical system governed by fractional order is the one by Caputo. For defining the equilibrium point consider the following Caputo fractional order autonomous system

$$ {}_{t_0}^{C}D_t^{\alpha} x(t) = f(t, x), \tag{2.41}$$

with initial condition $x(t_0)$, where $\alpha \in (0\ \ 1)$, $f : [t_o, \infty] \times \boldsymbol{\Omega} \to \mathbb{R}^n$ is piecewise continuous in t and locally Lipschitz in x on $[t_o, \infty] \times \boldsymbol{\Omega}$, $\boldsymbol{\Omega} \in \mathbb{R}^n$ is a domain that contains the equilibrium point $x = 0$.

Definition 2.8 The constant x_0 is an equilibrium point of the Caputo fractional dynamic system (2.41), if and only if $f(t, x_0) = 0$.

Remark 2.9 When $\alpha \in (0 \;\; 1)$, the Caputo fractional order system (2.41) has the same equilibrium points as the integer-order system $\dot{x}(t) = f(t, x)$.

After defining the equilibrium point, the most fundamental aspect about any dynamical system is the stability of system with respect to the equilibrium point. So in the next subsection we review some of the fundamental definitions of stability. It can be seen that these concepts are similar to those of integer order.

2.3.2 Fundamental of Stability

Definition 2.10 [5] The zero solution of ${}_0D_t^{\alpha}x(t) = f(t, x)$ is said to be stable if, for any initial conditions $x(0) \in \mathbb{R}^n$, there exists $\delta > 0$ such that any solution $x(t)$ of ${}_0D_t^{\alpha}x(t) = f(t, x)$ satisfies $\|x(t)\| < \delta$ for all $t > t_0$. Further, the zero solution of fractional differential system is said to be asymptotically stable if, in addition to being stable, $\|x(t)\| \to 0$ as $t \to \infty$.

Similarly $L^p(\boldsymbol{\Omega})$ stability of fractional order system with Riemann-Liouville derivative is defined as follows:

Definition 2.11 Suppose that $1 \leq p \leq \infty$ and $\boldsymbol{\Omega} \subset [t_0, \infty]$, then the solution $x(t)$ of the fractional order differential system ${}_0D_t^{\alpha}x(t) = f(t, x)$ (where $0 < \alpha < 1$, $x(0) \in \mathbb{R}^n$ is the initial condition , $f \in \mathbb{C}([t_0, \infty)$ is a continuous positive function) is called $L^p(\boldsymbol{\Omega})$ stability if

$$x(t) = \frac{x(0)}{\Gamma(\alpha)}(t - t_0)^{\alpha-1} + \frac{1}{\Gamma(\alpha)}\int_{t_0}^{t}(t - \tau)^{\alpha-1}f(\tau, x(\tau))d\tau,$$

belongs to $L^p(\boldsymbol{\Omega})$.

Other more generalized stability concept similar to asymptotic stability will be discussed in the next subsection. It is found that it is more convenient to characterize stability according to linear and non linear fractional order systems.

2.3.3 $t^{-\alpha}$ Stability

Decay rate of a simple fractional order autonomous system is not the same as of integer order. The components of the state variables in fractional order system have anomalous decay, due to the fact that fractional order systems have memory features. Asymptotic stability is also called $t^{-\alpha}$ stability. Following definition has been suggested by Sabatier et al. (2010) [6] regarding this proposition.

Definition 2.12 The trajectories $x(t) = 0$ of the system $d^{\alpha}x(t)/dt^{\alpha} = f(t, x(t))$ is $t^{-\alpha}$ asymptotic stable if the system is uniformly asymptotically stable and if there is a positive real α such that:

$$\begin{aligned}&\forall\ \|x(t)\|,\ t \le t_0\ \ \exists\, N\ (x(t),\ t \le t_0),\ t_1(x(t),\ t \le t_0)\\&\text{such that } \forall\, t > t_0\ \ \|x(t)\| \le N\ (t - t_1)^{-\alpha}.\end{aligned} \tag{2.42}$$

2.3.4 Mittag-Leffler Stability

Before defining the Mittag-Leffler stability, the following concepts are primarily required.

Lemma 2.13 [4] *Fractional integral of real valued function $f(t, x)$ satisfies the following inequality*

$$\|{}_{t_0}D_t^{-\alpha} f(t, x(t))\| \le {}_{t_0}D_t^{-\alpha} \|f(t, x(t))\|, \tag{2.43}$$

where $\alpha \ge 0$ and $\|.\|$ denotes the arbitrary norm.

Proof Taking the arbitrary norm on the fractional order integral, one can write

$$\begin{aligned}\|{}_{t_0}D_t^{-\alpha} f(t, x(t))\| &= \left\| \frac{1}{\Gamma(\alpha)} \int_{t_0}^{t} \frac{f(\tau, x(\tau))}{(t-\tau)^{1-\alpha}} d\tau \right\|\\&\le \frac{1}{\Gamma(\alpha)} \int_{t_0}^{t} \frac{\|f(\tau, x(\tau))\|}{(t-\tau)^{1-\alpha}} d\tau\\&\le {}_{t_0}D_t^{-\alpha} \|f(t, x(t))\|.\end{aligned} \tag{2.44}$$

This ends the proof.

Theorem 2.14 [4] *If $x = 0$ is the equilibrium point of the system ${}_{t_0}^{C}D_t^{\alpha} x(t) = f(t, x)$, f is Lipschitz by a constant L and is piecewise continuous with respect to t, then the solution of the system satisfies $\|x(t)\| \le \|x(t_0)\|\ E_{\alpha}(L(t - t_0)^{\alpha})$, where $\alpha \in (0, 1)$.*

Proof Applying ${}_{t_0}D_t^{-\alpha} f(t, x(t))$ to both side of ${}_{t_0}^{C}D_t^{\alpha} x(t) = f(t, x)$, one can write

$$x(t) = x(t_0) + {}_{t_0}D_t^{-\alpha} f(t, x(t)). \tag{2.45}$$

Using norm-inequality, the above equation can be rewritten as,

$$\|x(t)\| - \|x(t_0)\| \le \|x(t) - x(t_0)\| \le \left\| {}_{t_0}D_t^{-\alpha} f(t, x(t)) \right\|. \tag{2.46}$$

Further, using Lipschitz condition and Lemma 2.13, one can write

$$\left\|{}_{t0}D_t^{-\alpha} f(t, x(t))\right\| \leq {}_{t0}D_t^{-\alpha} \left\|f(t, x(t))\right\| \leq L\, {}_{t0}D_t^{-\alpha} \left\|x(t)\right\|. \tag{2.47}$$

There exists a nonnegative function $\eta(t)$ satisfying

$$\|x(t)\| - \|x(t_0)\| = L\, {}_{t0}D_t^{-\alpha} \|x(t)\| - \eta(t). \tag{2.48}$$

By applying Laplace transform to (2.48), one can further write

$$\|x(t)\| = \frac{\|x(t_0)\| s^{\alpha-1} - s^{\alpha}\eta(s)}{s^{\alpha} - L}. \tag{2.49}$$

Now applying Inverse Laplace transform to (2.49), which gives

$$\|x(t)\| = \|x(t_0)\| E_{\alpha}(L(t-t_0)^{\alpha}) - \eta(t) * \left[t^{-1} E_{\alpha,0}(L(t-t_0)^{\alpha})\right], \tag{2.50}$$

where $*$ denotes the convolution operator and

$$t^{-1} E_{\alpha,0}(L(t-t_0)^{\alpha}) = \frac{dE_{\alpha}(L(t-t_0)^{\alpha})}{dt} \geq 0. \tag{2.51}$$

Using (2.50) and (2.51), one can write

$$\|x(t)\| \leq \|x(t_0)\| E_{\alpha}(L(t-t_0)^{\alpha}). \tag{2.52}$$

This ends the proof.

The following Lemma is important for establishing the relation between Riemann-Lioville and Caputo fractional derivative

Lemma 2.15 *Suppose that* $\alpha \in (0, 1)$ *and* $f(0) \geq 0$ *then*

$${}_{t0}^{C}D_t^{\alpha} f(t) \leq {}_{t0}^{RL}D_t^{\alpha} f(t). \tag{2.53}$$

Proof One can write

$${}_{t0}^{C}D_t^{\alpha} f(t) = {}_{t0}^{RL}D_t^{\alpha-1} \frac{d}{dt} f(t) = {}_{t0}^{RL}D_t^{\alpha} f(t) - \frac{f(0)t^{-\alpha}}{\Gamma(1-\alpha)}. \tag{2.54}$$

Now substituting the condition $\alpha \in (0, 1)$ and $f(0) \geq 0$, then ${}_{t0}^{C}D_t^{\alpha} f(t) \leq {}_{t0}^{RL}D_t^{\alpha} f(t)$. This ends the proof.

Based on Theorem 2.14, the following stability condition has been proposed by Li et al. (2009) [4], which has been termed as Mittag-Leffler stability.

Definition 2.16 The solution of ${}_{t_0}D_t^{\alpha}x(t) = f(t, x)$ is said to be Mittag-Leffler stable if

$$||x(t)|| \leq \{m[x(t_0)]E_{\alpha}(-\lambda(t - t_0)^{\alpha})\}^{b} \tag{2.55}$$

where $E_{\alpha}(z) = \sum_{k=0}^{\infty} \frac{z^k}{\Gamma(k\alpha+1)}$, t_0 is the initial time $\alpha \in (0, 1)$, $\lambda > 0$, $b > 0$, $m(0) = 0$, $m(x) \geq 0$, and $m(x)$ is locally Lipschitz on $x \in \mathbb{B} \in \mathbb{R}^n$ with Lipschitz constant m_0.

Remark 2.17 Mittag-Leffler stability implies asymptotic stability.

2.3.5 Stability Using Ω Plane Analysis

For the simplicity let us assume that the following fractional order differential equation [7]

$${}_0D_t^{\alpha}x(t) = -ax(t) + bu(t), \tag{2.56}$$

where $x(t) \in \mathbb{R}$ and $u(t) \in \mathbb{R}$ is the control input. It is assumed that all initial conditions, or initialization functions, are zero. Then the Laplace transform of (2.56) is given as

$$s^{\alpha}X(s) = -aX(s) + bU(s). \tag{2.57}$$

System transfer of (2.56) is given as

$$G(s) = \frac{X(s)}{U(s)} = \frac{b}{s^{\alpha} + a}. \tag{2.58}$$

Impulse response of (2.58)

As b is the constant, it can be assumed to be unity without any loss of generality. Now expanding the right hand side of (2.58) about $s = \infty$, one can write

$$G(s) = \frac{1}{s^{\alpha} + a} = \frac{1}{s^{\alpha}} - \frac{a}{s^{2\alpha}} + \frac{a^2}{s^{3\alpha}} - \cdots = \frac{1}{s^{\alpha}} \sum_{j=0}^{\infty} \frac{(-a)^j}{s^{j\alpha}}. \tag{2.59}$$

The inverse Laplace transform of (2.59), using the fact $\frac{1}{s^{\alpha}} = \mathscr{L}\left\{\frac{t^{\alpha-1}}{\Gamma(\alpha)}\right\}$

$$\begin{aligned} g(t) &= \mathscr{L}^{-1}\left\{\frac{1}{s^{\alpha}} - \frac{a}{s^{2\alpha}} + \frac{a^2}{s^{3\alpha}} - \cdots\right\} \\ &= \frac{t^{\alpha-1}}{\Gamma(\alpha)} - \frac{at^{2\alpha-1}}{\Gamma(2\alpha)} + \frac{a^2t^{3\alpha-1}}{\Gamma(3\alpha)} + \cdots . \end{aligned} \tag{2.60}$$

Hence generalized impulse response of function is given as

$$g(t) = t^{\alpha-1}\sum_{j=0}^{\infty}\frac{(-a)^j t^{j\alpha}}{\Gamma(j\alpha+\alpha)}. \tag{2.61}$$

Unit step response of (2.58)

If the input function $u(t)$ is a unit step function, (2.58) can be written as

$$X(s) = \frac{1}{s}\left[\frac{1}{s^\alpha + a}\right]. \tag{2.62}$$

Further, after rearranging (2.62)

$$X(s) = \frac{1/a}{s}\left[\frac{a}{s^\alpha + a}\right] = \frac{1/a}{s}\left[1 - \frac{s^\alpha}{s^\alpha + a}\right] = \frac{1/a}{s} - \frac{s^\alpha/a}{s(s^\alpha + a)}. \tag{2.63}$$

For obtaining Inverse Laplace transform of the above equation, the following definition of Mittage Leffler function in summation form is needed

$$E_\alpha(x) = \sum_{k=0}^{\infty}\frac{x^k}{\Gamma(k\alpha+1)}, \quad \alpha > 0. \tag{2.64}$$

Assuming $x = -at^\alpha$, (2.64) becomes

$$E_\alpha(-at^\alpha) = \sum_{k=0}^{\infty}\frac{(-a)^k t^{k\alpha}}{\Gamma(k\alpha+1)}, \quad \alpha > 0. \tag{2.65}$$

Taking Laplace transform of (2.65)

$$\begin{aligned}\mathscr{L}\left\{E_\alpha(-at^\alpha)\right\} &= \mathscr{L}\left\{\frac{1}{\Gamma(1)} - \frac{at^\alpha}{\Gamma(1+\alpha)} + \frac{a^2t^{2\alpha}}{\Gamma(1+2\alpha)} + \cdots\right\} \\ &= \frac{1}{s} - \frac{a}{s^{\alpha+1}} + \frac{a^2}{s^{2\alpha+1}} + \cdots,\end{aligned} \tag{2.66}$$

or, equivalently

$$\begin{aligned}\mathscr{L}\left\{E_\alpha(-at^\alpha)\right\} &= \frac{1}{s}\left[1 - \frac{a}{s^\alpha} + \frac{a^2}{s^{2\alpha}} + \cdots\right] = \frac{1}{s}\sum_{j=0}^{\infty}\left(\frac{-a}{s^\alpha}\right)^j \\ &= \frac{1}{s}\left[\frac{s^\alpha}{s^\alpha + a}\right].\end{aligned} \tag{2.67}$$

Using (2.67), the step response of the system can be obtained by taking the inverse Laplace transform of (2.63), which is given as

$$x(t) = \frac{1}{a}\left[H(t) - E_{\alpha}(-at^{\alpha})\right], \quad (2.68)$$

where $H(t)$ is the Heaviside unit step function.

Stability using (2.58)

Perform the following conformal transformation of s

$$\boldsymbol{\Omega} = s^{\alpha}. \quad (2.69)$$

Then (2.58) is transformed as

$$G(s) = \frac{b}{s^{\alpha} + a} \Leftrightarrow \frac{1}{\boldsymbol{\Omega} + a}. \quad (2.70)$$

Using the above transformation, we will study the $\boldsymbol{\Omega}$-plane poles. Once the time domain responses are obtained corresponding to the $\boldsymbol{\Omega}$-plane pole locations, their behavior in new complex plane can be characterized.

For this, it is necessary to map the s-plane, along with the time-domain function properties associated with each point, into the new complex $\boldsymbol{\Omega}$-plane. For simplicity assume that $0 < \alpha \leq 1$. Then (2.69) can be written as

$$\boldsymbol{\Omega} = s^{\alpha} = \left(re^{j\theta}\right)^{\alpha} = r^{\alpha}e^{j\alpha\theta}. \quad (2.71)$$

Using (2.71), it is possible to map s-plane into the $\boldsymbol{\Omega}$-plane. For the stability, mapping of imaginary axis $s = re^{\pm j\pi/2}$ is important. The image of this axis in the $\boldsymbol{\Omega}$-plane is

$$\boldsymbol{\Omega} = r^{\alpha}e^{\pm j\frac{\alpha\pi}{2}}, \quad (2.72)$$

which is the pair of lines at $\phi = \frac{\pm\alpha\pi}{2}$, where ϕ is the angle in $\boldsymbol{\Omega}$ plane and $\boldsymbol{\Omega} = \rho e^{j\phi}$. Thus, the right half of the s-plane maps into a wedge in the $\boldsymbol{\Omega}$-plane of angle less than $\pm\frac{\pi}{2}\alpha$ degrees, that is the right half s-plane maps into $|\phi| < \frac{\alpha\pi}{2}$, which is shown in Fig. 2.1. Similar, kind of situation in the case of $1 < \alpha \leq 2$ is which shown in Fig. 2.2.

Example Inductor terminated semi-infinite lossy line

Consider the system shown in Fig. 2.3, where the inductor is terminated on the lossy line. The input to the system is voltage $v_i(t)$ and output $v_0(t)$ will be selected as the terminal of the lossy line. Assume that the $L = 1$, then transfer function is expressed as [7]

$$G(s) = \frac{V_O(s)}{V_I(s)} = \frac{\frac{1}{\sqrt{s}}}{s + \frac{1}{\sqrt{s}}} = \frac{1}{s^{3/2} + 1}. \tag{2.73}$$

Taking the inverse Laplace transformation, the above problem can be expressed in time domain as

$${}_0D_t^{\frac{3}{2}} v_O(t) + v_O(t) = v_i(t), \tag{2.74}$$

where all the initial conditions are assumed to be zero. One can also note the following:

- Impulse response of the system is given as (2.61) by substituting $\alpha = 3/2$

$$v_O(t) = \mathscr{L}^{-1}\left\{\frac{1}{s^{3/2}+1}\right\} = t^{1/2}\sum_{j=0}^{\infty}\frac{(-1)^j t^{3/2j}}{\Gamma(j3/2+3/2)}. \tag{2.75}$$

- The step response of the system is given as (2.68) by substituting $\alpha = 3/2$

$$v_O(t) = \mathscr{L}^{-1}\left\{\frac{1}{s(s^{3/2}+1)}\right\} = H(t) - E_{3/2}[-t^{3/2}]. \tag{2.76}$$

- For the stability of (2.73), let the transformation is taken as $s^{1/2} = \boldsymbol{\Omega}$. Then the transfer function is given as

$$G(s) = \frac{V_O(s)}{V_I(s)} = \frac{1}{\boldsymbol{\Omega}^3 + 1}. \tag{2.77}$$

Fig. 2.1 $0 < \alpha < 1$

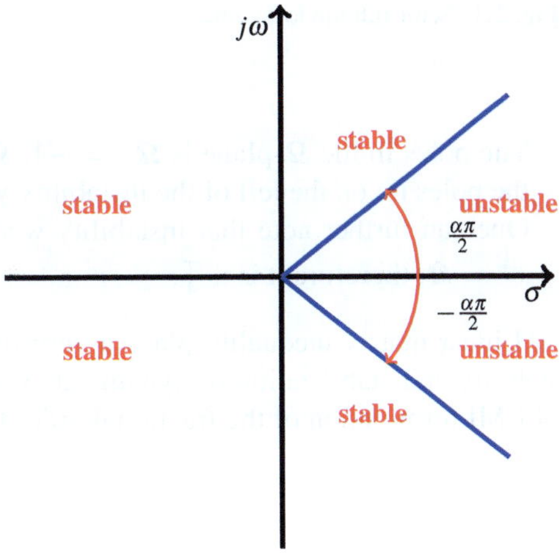

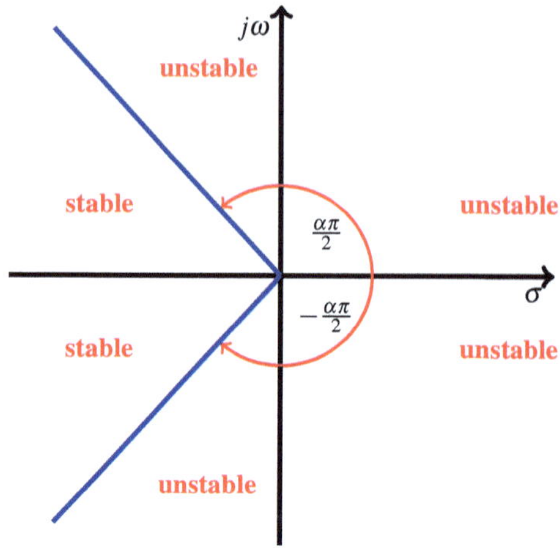

Fig. 2.2 $1 < \alpha < 2$

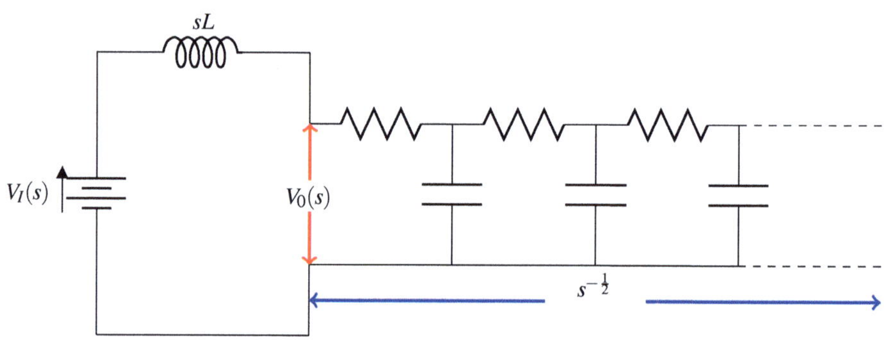

Fig. 2.3 Semi-infinite lossy line

The poles in the $\boldsymbol{\Omega}$-plane is $\boldsymbol{\Omega}_1 = -1$, $\boldsymbol{\Omega}_2 = e^{+\frac{j\pi}{3}}$ and $\boldsymbol{\Omega}_3 = e^{-\frac{j\pi}{3}}$. Hence, all the poles lie on the left of the instability wedge $\phi = \pm\frac{\pi}{4}$ and the system is stable. One can further note that instability wedge is calculated based on the mapping $s^{\frac{1}{2}} = \boldsymbol{\Omega}$, therefore it is $\pm\frac{\pi}{4}$.

Linear matrix inequality plays a very important role in control theory for both stability and stabilization of dynamical systems. In the next section, a brief review of LMI formulation of the fractional order linear systems is given.

2.4 A Brief Review on Linear Matrix Inequality (LMI) Stability Conditions for LTI Fractional Order Systems

State space representation [8] of a fractional order linear time-invariant system is given as

$$\begin{aligned} {}_0D_t^{\alpha}x(t) &= Ax(t) + Bu(t),\ x(0) = x_0 \\ y(t) &= Cx(t) \end{aligned} \tag{2.78}$$

where $x(t) \in \mathbb{R}^n$, $u(t) \in \mathbb{R}^r$ and $y(t) \in \mathbb{R}^p$ are states, input and output vectors of the system and $A \in \mathbb{R}^{n\times n}$, $B \in \mathbb{R}^{n\times r}$, $C \in \mathbb{R}^{p\times n}$ and $0 < \alpha < 2$ is the fractional commensurate order and pair (A, B) is controllable.

It has been well established in literature that the controllability and observability conditions of the continuous-time commensurate fractional order systems are same as that of the integer order case [9]. Thus, the system (5.1) is controllable if the rank of the controllability matrix

$$\mathbb{C} = \begin{bmatrix} B & AB & A^2B & \cdots & A^{n-1}B \end{bmatrix}, \tag{2.79}$$

is equal to n. Similarly, the system (5.1) is observable if the rank of the observability matrix

$$\mathbb{O} = \begin{bmatrix} C \\ CA \\ CA^2 \\ \vdots \\ CA^{n-1} \end{bmatrix}. \tag{2.80}$$

Theorem 2.18 [6, 10] *The system ${}_0D_t^{\alpha}x(t) = Ax(t)$ is asymptotically stable if the following condition is satisfied*

$$|arg(eig(A))| > \frac{\alpha\pi}{2}, \tag{2.81}$$

where $0 < \alpha < 2$ and $eig(A)$ are eigenvalues of matrix A.

Based on the mapping in $\boldsymbol{\Omega}$ plane the following theorem has been proved for the stability of linear time invariant fractional order system

Theorem 2.19 [11, 12] *The fractional order linear time invariant system ${}_0D_t^{\alpha}x(t) = Ax(t)$ where $1 < \alpha < 2$, is $t^{-\alpha}$ stable if and only if there exist a positive definite Hermitian matrix $P = P^* > 0$ such that*

$$\beta PA + \beta^* A^T P < 0, \tag{2.82}$$

where $\beta = \eta + j\zeta$ *and* η, ζ *are defined from* $\tan(\frac{\pi}{2} - \theta) = \frac{\eta}{\zeta}$ *with* $\theta = (\alpha - 1)\frac{\pi}{2}$.

Based on the above Theorem, the following lemma is recently derived.

Lemma 2.20 [13] *The fractional order linear time invariant system* ${}_0D_t^\alpha x(t) = Ax(t)$ *where* $1 < \alpha < 2$, *is stable (regarding input and output) if and only if the following integer order system is stable*

$$\dot{x}(t) = (\beta A)x(t). \tag{2.83}$$

where $\beta = \eta + j\zeta$ *and* η, ζ *are defined from* $\tan(\frac{\pi}{2} - \theta) = \frac{\eta}{\zeta}$ *and* $\theta = (\alpha - 1)\frac{\pi}{2}$.

Remark 2.21 The above lemma establishes the relationship of the LMI inequality (2.82) with an integer order linear system (2.83) which ensures the stability of the linear time invariant fractional order system ${}_0D_t^\alpha x(t) = Ax(t)$. Therefore (2.83) is nothing but a shadow (equivalent) system of ${}_0D_t^\alpha x(t) = Ax(t)$ from the stability point of view.

The above mentioned stability condition can be used to illustrate the state feedback control design for the fractional system (5.1). Let us consider a stabilizing control of the form $u = Kx$. The closed loop system becomes

$${}_0D_t^\alpha x(t) = (A + BK)x(t) \tag{2.84}$$

The necessary and sufficient condition for stability of system (2.84), according to Lemma (2.20) and Theorem (2.19),

$$\beta P(A + BK) + \beta^*(A + BK)^T P \leq -R \;\; \forall x \in \mathbb{R}^n, \tag{2.85}$$

where P and R are the symmetric positive definite matrices.

Following remark is necessary for checking the negative definiteness of complex Hermitian matrix H.

Remark 2.22 [11] A complex Hermitian matrix H is negative definite($H < 0$), if and only if

$$\begin{bmatrix} \mathrm{Re}(H) & \mathrm{Im}(H) \\ -\mathrm{Im}(H) & \mathrm{Re}(H) \end{bmatrix} < 0, \tag{2.86}$$

where $\mathrm{Re}(H)$ and $\mathrm{Im}(H)$ are the real and imaginary part of Hermitian matrix H respectively.

Theorem 2.23 [11, 12] *The fractional order linear time invariant system* ${}_0D_t^\alpha x(t) = Ax(t)$ *where* $1 < \alpha < 2$, *is* $t^{-\alpha}$ *asymptotically stable if and only if there exist a positive symmetric definite matrix* $P = P^\top > 0$, $P \in \mathbb{R}^{n \times n}$, *such that*

$$\begin{bmatrix} (A^\top P + PA)\sin(\alpha\frac{\pi}{2}) & (A^\top P - PA)\cos(\alpha\frac{\pi}{2}) \\ (PA - A^\top P)\cos(\alpha\frac{\pi}{2}) & (A^\top P + PA)\sin(\alpha\frac{\pi}{2}) \end{bmatrix} < 0. \tag{2.87}$$

Proof The system ${}_0D_t^{\alpha} = Ax(t)$ is $t^{-\alpha}$ asymptotically stable if the following condition is satisfied

$$|\arg(\lambda)| > \frac{\alpha\pi}{2},$$

where $0 < \alpha < 2$ and λ are eigenvalues of the matrix A. Now define following regions

- rotate λ by an angles $(\alpha - 1)\frac{\pi}{2}$

$$R1 = \left\{\lambda \in \mathbb{C} : \text{Re}\left(\lambda e^{j(\alpha-1)\frac{\pi}{2}}\right) < 0\right\}, \tag{2.88}$$

  and
- rotate λ by an angles $(1 - \alpha)\frac{\pi}{2}$

$$R2 = \left\{\lambda \in \mathbb{C} : \text{Re}\left(\lambda e^{j(1-\alpha)\frac{\pi}{2}}\right) < 0\right\}, \tag{2.89}$$

where λ belongs to $R = R1 + R2$, $\mathbb{C}$: represents complex number and Re: represents real part. As for some $\lambda \in \text{spec}(A)$, conjugate $\lambda^* \in \text{spec}(A)$, and as $R1$ and $R2$ are symmetric with respect to the real axis of the complex plane.

$$\exists\lambda_1 \in \text{spec}(A),\ \lambda_1 \in R1 \Leftrightarrow \exists\lambda_2 \in \text{spec}(A),\ \lambda_2 \in R2, \tag{2.90}$$

hence only relation (2.89) is necessary to derive the stability. Also, (2.89) can be written as

$$\begin{aligned} &\lambda e^{j(1-\alpha)\frac{\pi}{2}} + \lambda^* e^{-j(1-\alpha)\frac{\pi}{2}} < 0 \\ &\Rightarrow \lambda\left(\cos\left((1-\alpha)\frac{\pi}{2}\right) + j\sin\left((1-\alpha)\frac{\pi}{2}\right)\right) \\ &\quad + \lambda^*\left(\cos\left((\alpha-1)\frac{\pi}{2}\right) + j\sin\left((\alpha-1)\frac{\pi}{2}\right)\right) < 0, \end{aligned} \tag{2.91}$$

because for any complex number z, $\text{Re}(z) = \frac{z+z^*}{2}$. From, above relation is true (see Boyd 1994 [14]) if and only if $\exists P > 0,\ P \in \mathbb{R}^{n\times n}$ the following LMI is feasible

$$(A^\top P + PA)\sin\left(\alpha\frac{\pi}{2}\right) + j(A^\top P - PA)\cos\left(\alpha\frac{\pi}{2}\right) \le 0. \tag{2.92}$$

As an LMI involving real term can be derived from a complex one, the problem becomes:

$$\begin{bmatrix} (A^\top P + PA)\sin(\alpha\frac{\pi}{2}) & (A^\top P - PA)\cos(\alpha\frac{\pi}{2}) \\ (PA - A^\top P)\cos(\alpha\frac{\pi}{2}) & (A^\top P + PA)\sin(\alpha\frac{\pi}{2}) \end{bmatrix} < 0.$$

When $0 < \alpha < 1$, the stability domain is not convex. Due to the absence of the convexity property, the LMI conditions can not be derived directly as in the case of integer order or fractional order with $1 < \alpha < 2$. However, in literature different approaches are suggested to by pass this problem and LMI condition are derived indirectly. Some of the well recognized results which exist in literature are discussed here.

Theorem 2.24 *The fractional order linear time invariant system ${}_0D_t^{\alpha}x(t) = Ax(t)$ where $0 < \alpha < 1$, is $t^{-\alpha}$ asymptotically stable if and only if there exist positive definite Hermitian matrices $H_1 = H_1^* \in \mathbb{C}^{n\times n}$ and $H_2 = H_2^* \in \mathbb{C}^{n\times n}$ such that*

$$\bar{r}H_1A^\top + rAH_1 + rH_2A^\top + \bar{r}AH_2 < 0, \tag{2.93}$$

where $r = e^{j(1-\alpha)\frac{\pi}{2}}$.

Theorem 2.25 [15] *The fractional order linear time invariant system ${}_0D_t^{\alpha}x(t) = Ax(t)$ where $0 < \alpha < 1$, is asymptotically stable if and only if there exist two real positive symmetric definite matrices $P_{k1} \in \mathbb{R}^{n\times n}$, $k = 1, 2$, and two skew-symmetric matrices $P_{k2} \in \mathbb{R}^{n\times n}$, $k = 1, 2$ such that*

$$\sum_{i=1}^{2}\sum_{j=1}^{2} Sym\left\{\Theta_{ij} \otimes (AP_{ij})\right\} < 0, \tag{2.94}$$

$$\begin{bmatrix} P_{11} & P_{12} \\ -P_{12} & P_{11} \end{bmatrix} > 0, \quad \begin{bmatrix} P_{21} & P_{22} \\ -P_{22} & P_{21} \end{bmatrix} > 0, \tag{2.95}$$

where

$$\Theta_{11} = \begin{bmatrix} \sin\left(\alpha\frac{\pi}{2}\right) & -\cos\left(\alpha\frac{\pi}{2}\right) \\ \cos\left(\alpha\frac{\pi}{2}\right) & \sin\left(\alpha\frac{\pi}{2}\right) \end{bmatrix}, \quad \Theta_{12} = \begin{bmatrix} \cos\left(\alpha\frac{\pi}{2}\right) & \sin\left(\alpha\frac{\pi}{2}\right) \\ -\sin\left(\alpha\frac{\pi}{2}\right) & \cos\left(\alpha\frac{\pi}{2}\right) \end{bmatrix}$$
$$\Theta_{11} = \begin{bmatrix} \sin\left(\alpha\frac{\pi}{2}\right) & \cos\left(\alpha\frac{\pi}{2}\right) \\ -\cos\left(\alpha\frac{\pi}{2}\right) & \sin\left(\alpha\frac{\pi}{2}\right) \end{bmatrix}, \quad \Theta_{22} = \begin{bmatrix} -\cos\left(\alpha\frac{\pi}{2}\right) & \sin\left(\alpha\frac{\pi}{2}\right) \\ -\sin\left(\alpha\frac{\pi}{2}\right) & -\cos\left(\alpha\frac{\pi}{2}\right) \end{bmatrix}, \tag{2.96}$$

where $Sym\{X\}$ denotes the expression $X^\top + X$ and $\otimes$ is the Kronecker product of two matrices.

Proof Let us define $P_{k1} = \text{Re}(H_k)$, $P_{k2} = \text{Re}(H_k)$, $k = 1, 2$. Since $P_{k1} - jP_{k2}^\top = P_{k1} + jP_{k2}$, $e^{j\theta} = \cos(\theta) + j\sin(\theta)$ and $e^{-j\theta} = \cos(\theta) - j\sin(\theta)$, using Theorem 2.24 one can write

$$P_{11} + jP_{12} > 0, \quad P_{21} + jP_{22} > 0, \tag{2.97}$$

which is equivalent to (2.95) and

$$\begin{aligned}&(\cos\theta + j\sin\theta)(P_{11} - jP_{12}^{\top})A^{\top} + (\cos\theta - j\sin\theta)A(P_{11} + jP_{12})\\&+ (\cos\theta - j\sin\theta)(P_{21} - jP_{22}^{\top})A^{\top} + (\cos\theta + j\sin\theta)A(P_{21} + jP_{22})\\&< 0\end{aligned} \tag{2.98}$$

where $\theta = (1-\alpha)\frac{\pi}{2}$. Further, (2.98) can be written as

$$\begin{aligned}&\mathrm{Sym}\{AP_{11}\cos\theta + AP_{12}\sin\theta + AP_{12}\cos\theta - AP_{22}\sin\theta\} + j(P_{11}A^{\top} - AP_{11})\sin\theta\\&+ j(-P_{12}^{\top}A^{T} + AP_{12})\cos\theta + j(-P_{12}A^{-\top} + AP_{21})\sin\theta + j(-P_{22}^{\top}A^{\top}\\&+ AP_{22})\cos\theta < 0,\end{aligned} \tag{2.99}$$

which can be further written as

$$\begin{aligned}&\mathrm{Sym}\left\{AP_{11}\sin\left(\alpha\frac{\pi}{2}\right) + AP_{12}\cos\left(\alpha\frac{\pi}{2}\right)\right\} + \mathrm{Sym}\left\{AP_{21}\sin\left(\alpha\frac{\pi}{2}\right) - AP_{22}\cos\left(\alpha\frac{\pi}{2}\right)\right\}\\&+ j(P_{11}A^{\top} - AP_{11})\cos\left(\alpha\frac{\pi}{2}\right) + j(AP_{12} - P_{12}^{\top}A^{\top})\sin\left(\alpha\frac{\pi}{2}\right)\\&+ j(P_{21}A^{\top} - P_{21}A^{T})\cos\left(\alpha\frac{\pi}{2}\right) + j(AP_{22} - P_{22}^{\top}A^{\top})\sin\left(\alpha\frac{\pi}{2}\right) < 0,\end{aligned} \tag{2.100}$$

this is equivalent to (2.94). This ends the proof.

One more LMI based stability theorem for $0 < \alpha < 1$ is proposed by Sabatier et al.(2010) [6] based on the equivalence of fractional order system with integer order and by analyzing the geometric property of stability domain, which is stated as:

Theorem 2.26 *The fractional order linear time invariant system* ${}_0D_t^{\alpha}x(t) = Ax(t)$ *where* $0 < \alpha < 1$, *is asymptotically stable if and only if there exists a positive definite matrix* $P \in \mathbb{S}$, *where* $\mathbb{S}$ *denotes the set of symmetric matrices, such that*

$$\left(-(-A)^{\frac{1}{2-\alpha}}\right)^{\top} P + P\left(-(-A)^{\frac{1}{2-\alpha}}\right) < 0. \tag{2.101}$$

A Large class of the dynamical systems used for practical applications, are nonlinear in nature. So stability and stabilization of this class of the systems are also very important. In the next subsection a brief review of the stability of fractional order nonlinear system based on the second method of Lyapunov is discussed.

2.4.1 A Brief Review of the Stability of Nonlinear Fractional Order Systems Based on Lyapunov Function

It is well known fact that Lyapunov's second method provides a platform to analyze the stability of the system without solving explicitly the differential equations. This analysis is recently extended by the Li et al. [16] for the fractional order system also. While obtaining the solution of fractional order differential equation, it is seen that it contains more generalized function that is called Mittag-Leffler function, rather than the exponential function (in case of integer order differential equation). Therefore, more generalized stability concept, which is called Mittag-Leffler stability is defined for the fractional order. Following Theorems are reported in Li et al. [16].

Theorem 2.27 *Let $x = 0$ be an equilibrium point for the autonomous fractional-order system*

$$ {}_0D_t^{\alpha} x(t) = f(t, x), \tag{2.102}$$

and $\mathbb{D} \subset \mathbb{R}^n$ be a domain containing the origin. Assume that there exist a Lyapunov candidate $V(t, x(t)) : [0, \infty) \times \mathbb{D} \to \mathbb{R}$, which is a continuously differentiable function and locally Lipschitz with respect to x such that

$$\begin{aligned} \alpha_1 ||x||^a \le V(t, x) &\le \alpha_2 ||x||^{ab} \\ {}_0^C D_t^{\alpha} V(t, x) &\le -\alpha_3 ||x||^{ab}, \end{aligned} \tag{2.103}$$

where $t \ge 0$, $x \in \mathbb{D}$, $\alpha \in (0, 1)$, $\alpha_1, \alpha_2, \alpha_3$, α, a and b are arbitrary positive constants. Then $x = 0$ is Mittag-Leffler stable. If the assumptions hold globally on $\mathbb{R}^n$, then $x = 0$ is globally Mittag-Leffler stable.

Proof Using (2.103), one can write

$$ {}_0^C D_t^{\alpha} V(t, x) \le -\frac{\alpha_3}{\alpha_2} V(t, x). \tag{2.104}$$

It is always possible to find a function $\eta(t)$, such that

$$ {}_0^C D_t^{\alpha} V(t, x) + \eta(t) = -\frac{\alpha_3}{\alpha_2} V(t, x). \tag{2.105}$$

After taking Laplace transform,

$$\begin{aligned} s^{\alpha} V(s) - V(0)s^{\alpha-1} + \eta(s) &= -\frac{\alpha_3}{\alpha_2} V(s) \\ V(s) &= \frac{V(0)s^{\alpha-1} - \eta(s)}{s^{\alpha} + \frac{\alpha_3}{\alpha_2}}, \end{aligned} \tag{2.106}$$

where $V(0) = V(0, x(0))$ and $V(s) = \mathscr{L}\{V(t, x)\}$. If $x(0) = 0 \Rightarrow V(0) = 0$, the solution of (6.2) is $x = 0$. For the case when $x(0) \ne 0 \Rightarrow V(0) > 0$, the inverse

Laplace of (2.106) is given as

$$V(t) = V(0)E_\alpha\left(-\frac{\alpha_3}{\alpha_2}t^\alpha\right) - \eta(t) * \left[t^{\alpha-1}E_{\alpha,\alpha}\left(-\frac{\alpha_3}{\alpha_2}t^\alpha\right)\right]. \tag{2.107}$$

Since $t^{\alpha-1} > 0$ and $E_{\alpha,\alpha}\left(-\frac{\alpha_3}{\alpha_2}t^\alpha\right) > 0$, it follows from (2.107) that

$$V(t) \leq V(0)E_\alpha\left(-\frac{\alpha_3}{\alpha_2}t^\alpha\right). \tag{2.108}$$

After substitution of (2.108) into (2.103)

$$\|x(t)\| \leq \left[\frac{V(0)}{\alpha_1}E_\alpha\left(-\frac{\alpha_3}{\alpha_2}t^\alpha\right)\right]^{\frac{1}{a}}, \tag{2.109}$$

where $\frac{V(0)}{\alpha_1} > 0$ for $x(0) \neq 0$. Also, $x(0) = 0$ only when $\frac{V(0)}{\alpha_1} = 0$, because $V(t, x)$ is locally Lipschitz with respect to x. Further, using (2.109) it is concluded that (6.2) is Mittag-Leffler stable.

When system (6.2) is represented using Riemann-Liouville definition, then following inequality is required which is already discussed in (2.53)

$${}_{t_0}^{C}D_t^\alpha V(t) \leq {}_{t_0}^{RL}D_t^\alpha V(t) \leq -\alpha_3\|x\|^{ab}, \tag{2.110}$$

which further implies

$$\|x(t)\| \leq \left[\frac{V(0)}{\alpha_1}E_\alpha\left(-\frac{\alpha_3}{\alpha_2}t^\alpha\right)\right]^{\frac{1}{a}}.$$

This ends the proof.

Similar kind of proof can be extended for these two theorems also.

Theorem 2.28 [4] *Let $x = 0$ be an equilibrium point for (either Caputo or Riemann-Liouville) autonomous fractional-order system (6.2), where $f(t, x)$ satisfies the Lipschitz condition with Lipschitz constant $L > 0$ and $\alpha \in (0, 1)$. Assume that there exist a Lyapunov candidate $V(t, x(t))$ satisfying*

$$\begin{aligned} \alpha_1||x||^a \leq V(t, x) &\leq \alpha_2||x|| \\ \dot{V}(t, x) &\leq -\alpha_3||x||, \end{aligned} \tag{2.111}$$

where $\alpha_1, \alpha_2, \alpha_3$ and a are positive constants and $||.||$ denotes an arbitrary norm. Then the equilibrium point of the system (6.2) is Mittag-Leffer stable.

Theorem 2.29 [4] *Let $x = 0$ be an equilibrium point for the autonomous fractional-order system (6.2). Assume that there exists a Lyapunov function $V(t, x(t))$ and* *class-κ functions* $\alpha_1 (i = 1, 2, 3)$ *satisfying*

$$\alpha_1(||x||) \le V(t, x) \le \alpha_2(||x||)$$
$$ {}_0^C D_t^{\alpha} V(t, x) \le -\alpha_3(||x||). \tag{2.112}$$

Then the system (6.2) is asymptotically stable.

In general fractional calculus offers the following advantages to control engineering.

- adequate modeling of control plant's dynamic features
- effective robust control design
- reasonable realization by approximation.

A brief survey has been already presented in the initial part of the book about the first feature. Second feature is actually the main concern of this monograph and the last point is discussed in the next section.

2.5 Realization Issue of Fractional-Order Controller

It is obvious from the definition of fractional order operators that, to realize fractional order controllers perfectly, all the past input should be memorized. However, this is not possible in real scenario. Therefore, proper approximation by finite differential or difference equation must be introduced. There are many approximation methods exist in the literature. But, the most commonly used discretization method of a fractional-order controller is termed as *short memory principle* [2]. This discretization is based on the philosophy that, for the Grünwald-Letnikov definition, the values of the binomial coefficients near "*starting point*" $t = 0$ are small enough to be neglected or "forgotten" for large t. Therefore short memory principle takes into account the behavior of $f(t)$ only in "*recent past*", i.e., in the interval $[t - L, t]$, where L is the length of "*memory*"

$${}_{t_0}D_t^{\alpha} f(t) \approx {}_{t-L}D_t^{\alpha} f(t), \ t > t_0 + L, \ \alpha > 0. \tag{2.113}$$

Based on approximation of the time increment h through the sampling time T, the discrete equivalent of the fractional-order α derivative is given by

$$Z\left\{D^{\alpha} f(t)\right\} \approx \left(\frac{1}{T^{\alpha}} \sum_{j=0}^{m} c_j z^{-j}\right) F(z), \tag{2.114}$$

where $F(z) = Z\{f(t)\}$ $m = [L/T]$ and the coefficients c_j are

$$c_0 = 1$$

$$c_j = (-1)^j \binom{\alpha}{j} = \frac{j-\alpha-1}{j} c_{j-1}, \quad j \geq 1. \tag{2.115}$$

The following theorem is useful for showing the reasonability of the above approximation.

Theorem 2.30 *When the Riemann-Liouville (or the Grünwald-Letnikoff) definition is used, if* $|f(t)| < M, \forall t > t_0$, *then the error* ε *committed by approximation* (2.113) *is bounded by*

$$|\varepsilon| < \frac{M}{L^\alpha |\Gamma(1-\alpha)|}. \tag{2.116}$$

Proof When $\alpha < 0$,

$$|\varepsilon| < \left| {}_{t_0}D_t^\alpha f(t) - {}_{t-L}D_t^\alpha f(t) \right| = \left| {}_{t_0}D_{t-L}^\alpha f(t) \right|. \tag{2.117}$$

$$\begin{aligned} |\varepsilon| &\leq \left| \int_{t_0}^{t-L} \frac{(t-\tau)^{-\alpha-1}}{\Gamma(-\alpha)} f(\tau) d\tau \right| \leq \left| \int_{t_0}^{t-L} \frac{(t-\tau)^{-\alpha-1}}{\Gamma(-\alpha)} M d\tau \right| \\ &= \left| \frac{M}{\Gamma(-\alpha)} \left[\frac{-1}{\alpha} (t-\tau)^{-\alpha} \right]_{t_0}^{t-L} \right| = \left| \frac{M[(t-t_0)^{-\alpha} - L^{-\alpha}]}{\Gamma(1-\alpha)} \right|. \end{aligned} \tag{2.118}$$

Note This is not a very useful bound, because it grows with $t - t_0$ and can become very large. But if $\alpha > 0$

$$\begin{aligned} |\varepsilon| &\leq \left| \frac{d^{\lceil \alpha \rceil}}{dt^{\lceil \alpha \rceil}} {}_{t_0}D_t^{\alpha - \lceil \alpha \rceil} f(t) - \frac{d^{\lceil \alpha \rceil}}{dt^{\lceil \alpha \rceil}} {}_{t-L}D_t^{\alpha - \lceil \alpha \rceil} f(t) \right| \\ &= \left| \frac{d^{\lceil \alpha \rceil}}{dt^{\lceil \alpha \rceil}} {}_{t_0}D_{t-L}^{\alpha - \lceil \alpha \rceil} f(t) \right|. \end{aligned} \tag{2.119}$$

Using (2.118)

$$|\varepsilon| \leq \left| \frac{d^{\lceil \alpha \rceil}}{dt^{\lceil \alpha \rceil}} \frac{M[(t-t_0)^{-\alpha + \lceil \alpha \rceil} - L^{-\alpha + \lceil \alpha \rceil}]}{\Gamma(1 - \alpha + \lceil \alpha \rceil)} \right|. \tag{2.120}$$

Note that

$${}_0D_t^\alpha t^\lambda = \frac{\Gamma(\lambda+1)}{\Gamma(\lambda-\alpha+1)} t^{\lambda-\alpha}, \quad t \in \mathbb{R}^+, \quad \lambda \notin \mathbb{Z}^-. \tag{2.121}$$

Using (2.121), one can write

$$|\varepsilon| \leq \left| \frac{M(t-t_0)^{-\alpha}\Gamma(1-\alpha+\lceil\alpha\rceil)}{\Gamma(1-\alpha+\lceil\alpha\rceil-\lceil\alpha\rceil)\Gamma(1-\alpha+\lceil\alpha\rceil)} \right| = \frac{M}{(t-t_0)^{\alpha}|\Gamma(1-\alpha)|}. \tag{2.122}$$

Since $t - t_0 > 0$ which implies $(t-t_0)^\alpha > L^\alpha \Rightarrow \frac{1}{(t-t_0)^\alpha} < \frac{1}{L^\alpha}$

$$|\varepsilon| \leq \frac{M}{(t-t_0)^{\alpha}|\Gamma(1-\alpha)|} < \frac{M}{L^\alpha \Gamma(1-\alpha)}. \tag{2.123}$$

This ends the proof.

Remark 2.31 The above theorem is known as short memory principle, because (2.113) corresponds to a shortening of the memory of operator D, which remembers nothing older than L (thereby called memory length).

Corollary 2.32 *Thus to ensure the absolute value of error ε should not be larger than a certain value, the memory length in approximation* (2.113) *must satisfy*

$$L \geq \left(\frac{M}{|\varepsilon \Gamma(1-\alpha)|} \right)^{\frac{1}{\alpha}}, \quad \alpha > 0. \tag{2.124}$$

2.6 A Brief Review of Fractional Order PID Control

History of fractional order control started from the work of Bode [17, 18]. He formulated a problem to design a feedback amplifier to devise a feedback loop, so that the performance of the close-loop is invariant to changes in the amplifier gain. He gave a simple and elegant solution for this specified problem, which is termed as Bode's ideal loop transfer function, whose Nyquist plot is a straight line through the origin giving a phase margin invariant to gain changes. The Bode's ideal transfer function is represented as $G(s) = (\omega_0/s)^\alpha$, where $0 < \alpha < 1$, ω_0 is the gain crossover frequency and the constant phase margin is $\varphi_m = \pi - \frac{\alpha\pi}{2}$.

Above frequency characteristic is very interesting in term of robustness of the system to parameter changes or uncertainties. In fact, the fractional order integrator can be used as an alternative for more robust reference system for control. The frequency characteristics and the transient response of the non-integer order integral and its application to the control system was introduced by Manabe [19] and more recently by Barbosa et al. [20].

Analysis and design of controller for linear and nonlinear fractional order dynamical systems are easy and more efficient in time domain. Because, frequency domain approach is not easily extendable for the nonlinear or linear fractional order systems with disturbances which occurs in most of the cases. Several time design based controller design approaches are existing in the literature for the fractional order system as that for integer order system. One of the most successful controllers which is popular and useful in the practical industries is fractional order PID. The fractional order PID controller, namely the $PI^{\lambda}D^{\mu}$, which is the generalization of the classical PID controller is proposed by Podlubny [21] and Oustaloup [22–24]. In their series of papers and books Podlubny [2] and Oustaloup [22–24] successfully used the fractional order controller to develop the CRONE-controller (Commande Robuste d'Ordre Non Entrier controller), which is an interesting example of application of fractional calculus in control. He also demonstrated the superiority of fractional order $PI^{\lambda}D^{\mu}$ controller in comparison to the classical PID controller both for the fractional order and integer order dynamical systems.

Fractional order dynamical system can be represented in the time domain by the following differential equation

$$\left[\sum_{j=0}^{n} a_{n-j} D^{\alpha_{n-j}}\right] y(t) = f(t), \tag{2.125}$$

where $\alpha_{n-j} > \alpha_{n-j-1} (j = 0, 1, 2, \cdots, n) \in \mathbb{R}^{+}$, α_{n-j} are arbitrary constants, and $D^{\alpha} = {}_{0}^{C}D_{t}^{\alpha}$ denotes Caputo's fractional-order derivative of order α. The fractional-order transfer function for the system represented as (2.125) is given by

$$G_n(s) = \left[\sum_{j=0}^{n} a_{n-j} s^{\alpha_{n-j}}\right]^{-1}. \tag{2.126}$$

The unit-impulse response $y_i(t)$ of the system is given as follows

$$y_i(t) = \mathscr{L}^{-1}\{G_n(s)\} = g_n(t), \tag{2.127}$$

and the unit-step response function is given by the integral of the $g_n(t)$ so that

$$y_s(t) = \int_{0}^{t} g_n(\tau) d\tau. \tag{2.128}$$

2.6.1 *Brief Overview of Fractional Order Integral Action*

Following main effects are observed in case of integral actions

- it makes the system response slower
- decreases the system relative stability
- eliminates the steady state error for those inputs, for which the system had a finite error.

The effects of PID actions of a controller are analyzed using complex plane, time domain and frequncy domain methods. Similarly the fractional PID controller can be analyzed using these same techniques. For example consider the close loop system as shown in Fig. 2.4.

2.6.1.1 Complex Plane Analysis

In the complex plane, root locus of the system is displaced towards the right half plane after applying the integral action. Mathematically, the root locus of the system with control action is governed by

$$1 + Ks^{\alpha}G(s) = 0 \tag{2.129}$$

Its magnitude and phase is given as

$$\begin{aligned} |K| &= \frac{1}{|s^{\alpha}||G(s)|} \\ \arg\left[s^{\alpha}G(s)\right] &= (2n+1)\pi, \quad n = 0, \pm 1, \pm 2, \ldots. \end{aligned} \tag{2.130}$$

$s = |s|e^{j\theta}$ can be written as

$$s^{\alpha} = |s|^{\alpha}e^{j\alpha\theta}. \tag{2.131}$$

The conditions of phase can be further expressed as

$$\arg\left[s^{\alpha}G(s)\right] = \arg\left[G(s)\right] + \alpha\theta = (2n+1)\pi, \quad n = 0, \pm 1, \pm 2, \ldots. \tag{2.132}$$

Therefore, it is obvious that, by choosing $\alpha \in (-1, 0)$, the root locus is displaced towards the right half plane.

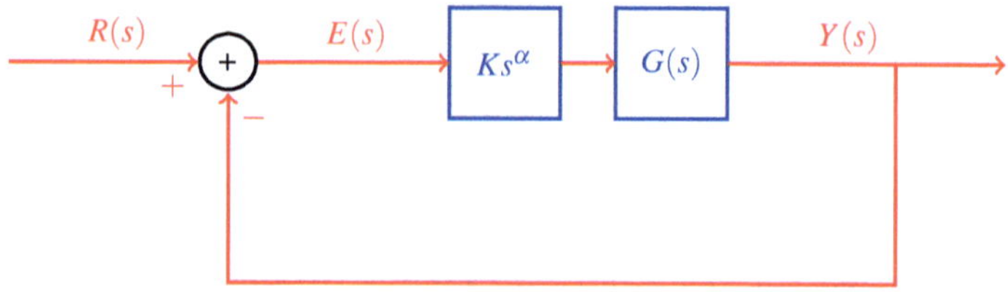

Fig. 2.4 Fractional integral action $\alpha \in (-1, 0)$

2.6.2 Frequency Domain Analysis

In frequency domain, a pole at zero adds $-20\,\text{dB/dec}$ in the magnitude curve and decreases the phase plot by $\pi/2$ rad. The effect of fractional order integral is explained as follows.

The magnitude curve in the frequency domain is given as

$$20\log\left[s^{\alpha}G(s)\right]_{s=jw} = 20\log|G(j\omega)| + 20\alpha\log\omega \tag{2.133}$$

and the phase plot is given by

$$\arg\left[s^{\alpha}G(s)\right]_{s=jw} = \arg|G(s)| + \alpha\frac{\pi}{2}. \tag{2.134}$$

Therefore, by varying the value of α between -1 and 0, it is possible to introduce a constant increment in the slope of the magnitude curve by introducing a fractional order integrator, which varies between -20 and 0 dB/dec. Similarly, a constant delay in phase plot, which varies between $-\frac{\pi}{2}$ and 0 rad.

2.6.3 Time Domain Analysis

By introducing a fractional order integrator, there are clear cut effects over the transient response, which consists of the decrease in the rise time, increase of the settling time and the overshoot. Mathematically, this effects can be studied considering the error signal of the following form

$$e(t) = \sum_{j=0}^{n}(-1)^{j}u_0(t - jT), \quad j = 0, 1, 2, \ldots, n \tag{2.135}$$

where $u_0(t)$ represents the unit step input. Its Laplace equivalent is given as

$$E(s) = \sum_{j=0}^{n}(-1)^{j}\frac{e^{-jTs}}{s}. \tag{2.136}$$

Therefore, the control action can be expressed as

$$\begin{aligned} u(t) &= \mathscr{L}^{-1}\{U(s)\} \\ &= \mathscr{L}^{-1}\left\{K\sum_{j=0}^{n}(-1)^{j}\frac{e^{-jTs}}{s^{1-\alpha}}\right\} \end{aligned}$$

$$= K \sum_{j=0}^{n} \frac{(-1)^j}{\Gamma(1-\alpha)} (t - jT)^{-\alpha} u_0 (t - jT) \quad (2.137)$$

It is clear that the control action over the error signal, vary between the effects of a proportional action $\alpha = 0$ (square signal) and an integral action $\alpha = -1$ (straight line curve). For the intermediate value of α, the control action increases for a constant error, which results in the elimination of the steady state error and decrease when error is zero, resulting a more stable system.

2.6.4 Brief Overview of Fractional Order Derivative Action

Same as fractional integral action, the derivative action of fractional order controller can be analyze in complex domain, frequency domain and time domain. For example consider the close loop system as shown in Fig. 2.5.

2.6.5 Complex Plane Analysis

In the complex plane, root locus of the system is displaced towards the left half plane after applying the derivative action.

2.6.6 Frequency Domain Analysis

In frequency domain, a derivative action of a controller adds a slope of +20 db/dec in the magnitude plot and adds $\pi/2$ radian in phase plot. Similarly in its fractional counter part, fractional derivative can add a slope of 0–20 db/dec, when β is varied. Similarly, a constant delay in phase plot, which varies between 0 and $\frac{\pi}{2}$ radian.

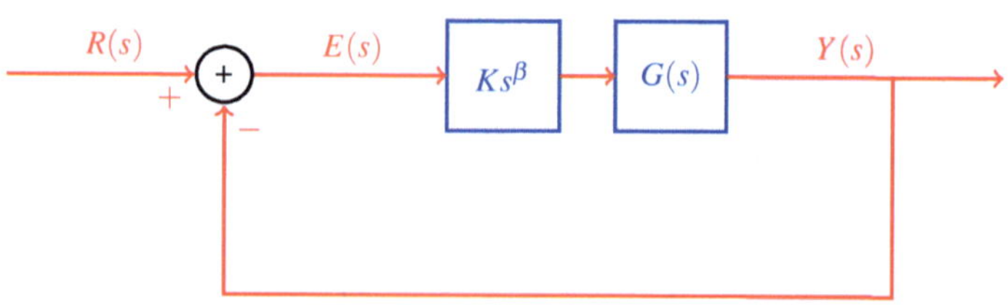

Fig. 2.5 Fractional derivative action $\beta \in (0, 1)$

2.6.7 Time Domain Analysis

In the time domain, a decrease in the overshoot and the settling time is observed. This can be studied using the trapezoidal error signal given as

$$e(t) = tu_0(t) - t(t-T)u_0(t-T) - t(t-2T)u_0(t-2T) + t(t-3T)u_0(t-3T), \tag{2.138}$$

where u_0 represents the unit step input. The laplace transform of (2.138) can be written as

$$E(s) = \frac{1}{s^2} - \frac{e^{-Ts}}{s^2} - \frac{e^{-2Ts}}{s^2} + \frac{e^{-3Ts}}{s^2}. \tag{2.139}$$

Therefore, the control action can be expressed as

$$\begin{aligned} u(t) &= \mathscr{L}^{-1}\{U(s)\} \\ &= \mathscr{L}^{-1}\left\{K\left(\frac{1}{s^{2-\beta}} - \frac{e^{-Ts}}{s^{2-\beta}} - \frac{e^{-2Ts}}{s^{2-\beta}} + -\frac{e^{-3Ts}}{s^{2-\beta}}\right)\right\} \\ &= \frac{K}{\Gamma(2-\beta)}\left\{t^{1-\beta}u_0(t) - (t-T)^{1-\beta}u_0(t-T) - (t-2T)^{1-\beta}u_0(t-2T)\right\} \\ &+ \frac{K}{\Gamma(2-\beta)}\left\{(t-3T)^{1-\beta}u_0(t-3T)\right\}. \end{aligned} \tag{2.140}$$

The effects of the fractional order control over the error signal vary between the effects of a proportional action $\beta = 0$ (trapezoidal signal) and a derivative action $\beta = 1$ (square signal).

2.6.8 The Fractional Order $PI^{\alpha}D^{\beta}$ Controller

The fractional order $PI^{\alpha}D^{\beta}$ controller is the generalization of the integer order PID controller. The transfer function of $PI^{\alpha}D^{\beta}$ controller is defined as the ratio of the controller output $U(s)$ and error $E(s)$ as

$$G(s) = \frac{U(s)}{E(s)} = K_P + K_I s^{-\alpha} + K_D s^{\beta}, \quad \alpha, \beta > 0. \tag{2.141}$$

Output $u(t) = \mathscr{L}^{-1}\{U(s)\}$, in the time domain as

$$u(t) = K_P e(t) + K_I D^{-\alpha} e(t) + K_D D^{\beta} e(t), \tag{2.142}$$

It is quite natural to conclude that by introducing more general control actions of the form $PI^{\alpha}D^{\beta}$, one could achieve more satisfactory performances between positive and negative effects of classical PID, and combining the fractional order actions one could develop more powerful and flexible design methods to satisfy the controlled system specifications.

2.6.9 Unit-Impulse and Unit-Step Response of the some Simple Transfer Function

Example Consider the following transfer function

$$G(s) = \left(a_0 s^{\alpha} + b_0\right)^{-1}, \quad \alpha > 0. \tag{2.143}$$

The unit-impulse and unit-step response of (2.143) are given as follows

$$y_i(t) = g(t) = \mathscr{L}^{-1}\left\{a_0 s^{\alpha} + b_0\right\}^{-1} = \frac{1}{a_0}\xi_0\left(t, -\frac{b_0}{a_0}; \alpha, \alpha\right). \tag{2.144}$$

$$y_s(t) = \int_0^t g(t) = \frac{1}{a_0}\xi_0\left(t, -\frac{b_0}{a_0}; \alpha, \alpha + 1\right). \tag{2.145}$$

where, $\xi_0\,(t, z; \alpha, \beta)$ is defined in terms of Mittage-Leffler's function $E_{\alpha,\beta}(x)$ as

$$\begin{aligned}
\xi_0\,(t, z; \alpha, \beta) &= t^{\alpha j+\beta-1}E^{(j)}_{\alpha,\beta}(zt^{\alpha}), \quad m = 0, 1, \ldots \\
E^{(j)}_{\alpha,\beta}(x) &= \frac{d^j}{dx^j}E_{\alpha,\beta}(x), \\
E_{\alpha,\beta}(x) &= \sum_{j=0}^{\infty}\frac{x^j}{\Gamma(\alpha j + \beta)}, \quad \alpha > 0,\ \beta > 0.
\end{aligned} \tag{2.146}$$

and the Laplace transfer is

$$\mathscr{L}\left\{t^{\alpha j+\beta-1}E^{(j)}_{\alpha,\beta}(\pm a_0 t^{\alpha})\right\} = \frac{j!s^{\alpha-\beta}}{(s^{\alpha} \mp a_0)^{m+1}}. \tag{2.147}$$

The function ξ_j satisfies the property

$${}_0D^n_t\xi_m\,(t, z; \alpha, \beta) = \xi_j\,(t, z; \alpha, \beta - n)\,, \quad \beta > n. \tag{2.148}$$

Example Consider the following transfer function

$$G(s) = \frac{1}{\left(a_0 s^{\alpha} + b_0 s^{\beta} + c_0\right)}, \quad \alpha > \beta > 0 \tag{2.149}$$

The fractional differential equation in the time domain is expressed as

$$a_0 y^{(\alpha)}(t) + b_0 y^{(\beta)}(t) + c y(t) = f(t) \tag{2.150}$$

with the following initial conditions

$$y(0) = y^{(1)}(0) = y^{(2)}(0) = 0. \tag{2.151}$$

The unit-impulse and unit-step response of (2.149) are given as follows

$$y_i(t) = g(t) = \mathscr{L}^{-1}\left\{G(s)\right\}, \tag{2.152}$$

$$y_s(t) = \int_0^t g(t) = \frac{1}{a_0}\sum_{j=0}^{\infty}\frac{(-1)^j}{j!}\left(\frac{c_0}{a_0}\right)^j \xi_j\left(t, \frac{b_0}{a_0}; \alpha-\beta, \alpha+\beta j+1\right). \tag{2.153}$$

2.7 Summary

This chapter discussed the solution issues of fractional differential equations and Mittag-Leffler function. Along this a brief summary of the notation of stability for the fractional order system are also presented. Linear matrix inequality (LMI) is the one of the most important tool to analyzed the stability and stabilization problem in control theory, this chapter present a brief review on that also. Ideal realization of the fractional order controller requires infinite memory, which is not practically feasible. Therefore, concept of short memory principle comes into picture. This issue is also discussed in this Chapter. Finally, fractional order PID and its benefits over classical PID control is presented.

References

1. Anatoly, A.K., Srivastava, H.M., Trujillo, J.J.: Theory and applications of fractional differential equations, vol. 204 (2006)
2. Podlubny, I.: Fractional differential equations: an introduction to fractional derivatives, fractional differential equations, to methods of their solution and some of their applications, vol. 198. Access Online via Elsevier (1998)

3. Du, M., Wang, Z.: Initialized fractional differential equations with riemann-liouville fractional-order derivative. Eur. Phys. J. Spec. Top. **193**(1), 49–60 (2011)
4. Li, Y., Chen, Y., Podlubny, I.: Mittag-leffler stability of fractional order nonlinear dynamic systems. Automatica **45**(8), 1965–1969 (2009)
5. Li, C., Zhang, F.: A survey on the stability of fractional differential equations. Eur. Phys. J. Spec. Top. **193**(1), 27–47 (2011)
6. Sabatier, J., Moze, M., Farges, C.: Lmi stability conditions for fractional order systems. Comput. Math. Appl. **59**(5), 1594–1609 (2010)
7. Hartley, T.T., Lorenzo, C.F.: A solution to the fundamental linear fractional order differential equation. NASA Lewis Research Center (1998)
8. Raynaud, H.F., Zergaınoh, A.: State-space representation for fractional order controllers. Automatica **36**(7), 1017–1021 (2000)
9. Chen, Y., Vinagre, B.M.: Fractional-Order Systems and Controls: Fundamentals and Applications. Springer, Berlin (2010)
10. Petráš, I.: Stability of fractional-order systems with rational orders: a survey. Fractional Calc. Appl. Anal. **12**(3), 269–298 (2009)
11. Ahn, H.S., Chen, Y., Podlubny, I.: Robust stability test of a class of linear time-invariant interval fractional-order system using lyapunov inequality. Appl. Math. Comput. **187**(1), 27–34 (2007)
12. Ahn, H.S., Chen, Y.: Necessary and sufficient stability condition of fractional-order interval linear systems. Automatica **44**(11), 2985–2988 (2008)
13. Balochian, S., Sedigh, A.K., Haeri, M.: Stabilization of fractional order systems using a finite number of state feedback laws. Nonlinear Dyn. **66**(1–2), 141–152 (2011)
14. Boyd, S.P., El Ghaoui, L., Feron, E., Balakrishnan, V.: Linear matrix inequalities in system and control theory, vol. 15. SIAM (1994)
15. Lu, J.G., Chen, Y.Q.: Robust stability and stabilization of fractional-order interval systems with the fractional order: The case. IEEE Trans. Autom. Control. **55**(1), 152–158 (2010)
16. Li, Y., Chen, Y., Podlubny, I.: Stability of fractional-order nonlinear dynamic systems: Lyapunov direct method and generalized mittag-leffler stability. Comput. Math. Appl. **59**(5), 1810–1821 (2010)
17. Bode, H.: Relations between attenuation and phase in feedback amplifier design. Bell Syst. Tech. J. **19**, 421–454 (1940)
18. Bode, H.W.: Network Analysis and Feedback Amplifier Design. Van Nostrand Reinhold, New York (1956)
19. Manabe, S.: The non-integer integral and its application to control systems. ETJ of Japan **6**(3–4), 83–87 (1961)
20. Barbosa, R.S., Machado, J.T., Ferreira, I.M.: A fractional calculus perspective of pid tuning. In: the Proceedings of the ASME International 19th Biennial Conference on Mechanical Vibration and Noise (VIB'03), Chicago, Illinois (2003)
21. Podlubny, I.: Fractional-order systems and pi/sup/spl lambda//d/sup/spl mu//-controllers. IEEE Trans. Autom. Control. **44**(1), 208–214 (1999)
22. Oustaloup, A.: From fractality to non integer derivation through recursivity, a property common to these two concepts: A fundamental idea for a new process control strategy. In: Proceedings of the 12th IMACS World Congress, vol. 3, pp. 203–208 (1988)
23. Oustaloup, A., Sabatier, J., Lanusse, P., et al.: From fractal robustness to crone control. Fractional Calc. Appl. Anal. **2**(1), 1–30 (1999)
24. Oustaloup, A., Melchior, P., Lanusse, P., Cois, O., Dancla, F.: The crone toolbox for matlab. In: Computer-Aided Control System Design, 2000. CACSD 2000. IEEE International Symposium on, pp. 190–195. IEEE (2000)

Chapter 3
Sliding Mode Control of Fractional Order Systems

3.1 Introduction

Robust design techniques are essential in any field of engineering design because the working and durability of their pieces of work is always jeopardized by mutable and unpredictable environments. Traditionally people are using safety factors to acknowledge the ignorance on the physical parameters variations and external influence of disturbance on design stage. However, in the last decades a lot of efforts are paid to both sophisticated mathematical modeling techniques, and advanced computational tools to capture more accurate model. Still a complete description of systems is impossible, or impracticable. For this reason in recent years robust control theory and application have received considerable attention and spread along several directions. There are several approaches in this direction for examples optimal control, H-infinity control, high gain feedback, variable structure etc. One of the novel ideas in this field is sliding mode control, which is subset of variable structure control.

The sliding mode control methodology is one such robust control technique which has its roots in the relay control. The research in this area was initiated in the former Soviet Union about 50 years ago, and the sliding mode control methodology has subsequently received much more attention from the international control community within the last two decades. One of the most intriguing aspects of sliding mode is the discontinuous nature of the control action whose primary function is to switch between two distinctively different structures about some predefined manifold such that a new type of system motion called sliding mode exists in a manifold. This peculiar system characteristic is claimed to result in a superb system performance which includes insensitivity to parameter variations and complete rejection of certain class of disturbances. Furthermore, the system possess new properties which are not present in original system.

Sliding mode control implies that control actions are discontinuous state functions which may be implemented by conventional power converters with 'on-off' as the only admissible operation mode or digital signal systems due to the increasing affordable microprocessor hardware. Due to these properties, the intensity of the research

B. Bandyopadhyay and S. Kamal, *Stabilization and Control of Fractional Order Systems: A Sliding Mode Approach*, Lecture Notes in Electrical Engineering 317,
DOI 10.1007/978-3-319-08621-7_3

at the many scientific centers of industry and universities is maintained at a high level, and sliding mode control has been proved to be applicable to a wide range of problems in robotics, electric drives and generators, process control, vehicle and motion control.

The idea of fractional calculus has been known since the development of the regular integer order calculus, with the first reference being associated with Leibniz and L'Hospital in 1695 where half order derivatives were mentioned. In the last two decades fractional differential equations have been used to model various stable physical phenomena. We can refer to [1] for the recent history of fractional calculus. Since it plays an important role in science and engineering, the study of stability of fractional differential equations has attracted much attention [2]. Furthermore, in recent years, an increasing attention has been given to fractional order controllers, and many progresses are reflected in [3, 4] and references cited therein. Some of these applications include optimal control [5], CRONE controller [6], fractional PID controller [7, 8], and lead-lag compensator [9].

Sliding mode controller [10] and VSC controller have been already implemented to improve the performance and robustness in the closed loop control systems. Sliding mode control (SMC) has many attractive features such as invariance to matched uncertainties, simplicity in design, robustness against perturbations and some others [10–12]. The characteristic feature of continuous-time SMC system is that sliding mode occurs on a prescribed manifold(sliding surface), where switching control is employed to maintain the states on the surface [13]. In the field of fractional order system control, the application of SMC was reported only recently in [14–16]. Asymptotic stabilization was achieved using higher order sliding mode control in [17]. However, the finite time stabilization of fractional order systems remains an unexplored field. To the best of authors' knowledge, robust finite time stabilization of fractional order system has not yet been achieved.

In this chapter firstly a brief introduction of existence of the solution of fraction order differential equation with discontinuous right-hand side is presented. Methodology of sliding surface and different aspects of the controller design for fractional order systems are also discussed. Finally we propose a methodology for finite time stabilization of fractional order systems using switched sliding surface algorithm. Since, sliding mode control is used, the proposed controller is robust to matched uncertainties. Moreover, the controller can be designed in such a way that the states reach the equilibrium point within some specified time.

The chapter has been organized in the following way: In Sect. 3.2, existence of the solution of fraction order differential equation with discontinuous right-hand side has been presented. Section 3.3 introduces the sliding surface design methodology for fractional order system. Fraction order differential equation with discontinuous right-hand side is presented in Sect. 3.4. A brief review on point to point control of fractional chain of integrator is discussed in Sect. 3.5. Proposed stabilization of uncertain fractional chain of integrator is presented in Sect. 3.6. In Sect. 3.7, the proposed control scheme is illustrated by numerical example. Finally, some concluding remarks are included in Sect. 3.8.

3.2 Existence of the Solution of Fraction Order Differential Equation with Discontinuous Right-Hand Side

A system of fractional order differential equation of order $\alpha > 0$ is given as

$$_0D_t^\alpha x(t) = f(x(t)). \tag{3.1}$$

The continuity of f is usually sufficient to ensure the existence of the solution of (3.1). Furthermore, when f fulfills a Lipschitz condition the uniqueness of the solution is also achieved. However, several control theoretical application the function f contains some discontinuities with respect to the state variable [18]. For example electrical systems (electrical circuits and networks with switches, power electronics), in mechanics (dry friction with stick and slip modes, impacts, oscillating systems with viscous damping, elasto-plasticity, alternatively forced vibrations, braking processes with locking phases), biological and physiological systems, also from automatic and optimal control theory, game theory, convex optimizations, non-smooth control systems synthesis, uncertain systems, walking machines, and everywhere non-smooth characteristics are used to represent switches. In other words, in the real world, non-smoothness is a common phenomenon.

However, when the vector field possesses only a piecewise smooth character the existence of the solution is no longer guaranteed. To overcome of the difficulties related to ODEs (Ordinary Differential Equation) with a non smooth character, Filippov [19] proposed. The regularization approach is discussed in the next subsection.

3.2.1 *A Brief Review of Filippov Theory*

In Filippov theory one seeks a solution of an initial value problem of ordinary differential equations in which the right-hand side (the vector field) varies discontinuously as the solution trajectory reaches one or more surfaces, called discontinuity or switching surfaces, but it is otherwise smooth; in the literature, these are called Piecewise Smooth Systems (PSW), hereafter PSW for short. The values where a trajectory reaches a discontinuity surface are called events, and we will henceforth assume that the events are isolated. In general, there are a number of possible outcomes as the solution reaches a discontinuity surface. For example, in the so-called impact systems, the solution experiences a jump discontinuity. However, in this discussion, we will consider only the case where the solution remains continuous (though not necessarily differentiable) past an event point. In this case, loosely speaking, there are two things which can occur as the trajectories reach a surface of discontinuity: it may cross the surface, or it may stay on the surface, in latter case a description of the motion on the surface will be required, which is called sliding motion.

Consider the nonlinear system with discontinuous right-hand side [20]

$$\dot{x}(t) = f(x(t)) = \begin{cases} f_1(x(t)) & \text{if } x(t) \in S_1 \\ f_2(x(t)) & \text{if } x(t) \in S_2 \end{cases} \quad x(0) = x_0 \in \mathbb{R}^n \tag{3.2}$$

The state space $\mathbb{R}^n$ is split into two subspaces S_1 and S_2 by a hypersurface Σ such that $\mathbb{R}^n = S_1 \bigcup \Sigma \bigcup S_2$. The hypersurface is defined by a scalar event function $h : \mathbb{R}^n \to \mathbb{R}$ (one dimensional manifold), so that the subspaces S_1 and S_2 and the hypersurface Σ are implicitly characterized as

$$\begin{aligned} \Sigma &= \left\{x \in \mathbb{R}^n \mid h(x) = 0\right\} \\ S_1 &= \left\{x \in \mathbb{R}^n \mid h(x) < 0\right\} \\ S_2 &= \left\{x \in \mathbb{R}^n \mid h(x) > 0\right\} \end{aligned} \tag{3.3}$$

We will assume that $h \in \mathbb{C}^k, k \geq 2$, and that $\nabla h(x) \neq 0$ for all $x \in \Sigma$. Thus, the unit normal n to Σ, perpendicular to the tangent plane $T_x(\Sigma)$ at $x \in \Sigma$, is given by

$$n(x) = \frac{\nabla(h(x))}{\|\nabla(h(x))\|}, \quad \|n(x)\| = 1 \text{ for all } x \in \Sigma \tag{3.4}$$

The right hand side of Eq. (3.2) can be assumed to be smooth on S_1 and S_2 separately, but it is usually discontinuous across Σ. Obviously $f(x(t))$ is not defined if $x(t)$ is on Σ. In the model (3.2), there is freedom to extend the vector field on Σ, and the way that this freedom is resolved must ultimately be weighed against our ability to model situations of practical interest, and to have mathematical background for the existence of solutions. A most important and widely accepted way to resolve this freedom is to consider the set valued extension $f(x(t))$ below

$$\dot{x}(t) \in f(x(t)) = \begin{cases} f_1(x(t)) & \text{if } x(t) \in S_1 \\ \bar{co}\{f_1(x(t)), f_2(x(t))\} & \text{if } x(t) \in \Sigma \\ f_2(x(t)) & \text{if } x(t) \in S_2 \end{cases} \tag{3.5}$$

where $\bar{co}(A)$ denotes the smallest closed convex set containing A. In our particular case,

$$\bar{co}\{f_1, f_2\} = \left\{f_F \in \mathbb{R}^n : f_F = (1-\alpha)f_1 + \alpha f_2, \alpha \in [0, 1]\right\}. \tag{3.6}$$

The extension (or convexification) of a discontinuous system (3.2) into a convex differential inclusion (3.6) is known as the Filippov convex method. Existence of solutions of (3.6) can be guaranteed with the notion of upper semicontinuity of set-valued function. More extensively [19]

Definition 3.1 A differential inclusion $\dot{x} \in F(x), x \in R^n$, is called a Filippov differential inclusion, if the vector set $F(x)$ is

1. Non-empty
2. Closed
3. Convex
4. Locally Bounded
5. Upper-semicontinuous

Such solutions always exists and have most of the well-known standard properties except the uniqueness.

Definition 3.2 It is said that a differential equation $\dot{x} = f(x), x \in R^n$, with a locally bounded Lebesgue- measurable right-hand side is understood in the Filippov sense, if it is replaced by a special Filippov differential inclusion $\dot{x} \in F(x)$, where

$$F(x) = \bigcap_{\delta>0} \bigcap_{\mu N=0} \overline{co} M\left(O_\delta(x)\ N\right). \tag{3.7}$$

Here μ is the Lebesgue measure, $O_\delta(x)$ is the δ-vicinity of x, and $\overline{co}M$ denotes the convex closure of M.

Note Just as a standard map or a function takes a point in its domain to a single point in another space, a set-valued map takes a point in its domain to a set of points in another space.

For the every locally bounded $f(x)$ its corresponding set-valued map $F(x)$ is upper semi-continuous, i.e., for any x in the domain of F and for any open set Z such that $F(x) \subset Z$ there exists a neighborhood X of x such that $F(X) \subset Z$, furthermore $F(x)$ assumes nonempty convex values and $F(x) = \{f(x)\}$ whenever f is continuous at x.

Filippov regularization approach can be extended to discontinuous fractional order (3.1) similar as integer order [21–23]. Solution for the problem (3.1) is any absolutely continuous map $x(t) : [0, T] \to \mathbb{R}^q$ such that the following differential inclusion

$${}_0D_t^\alpha x(t) \in F(x), \tag{3.8}$$

is satisfied for almost all $t \in [0, T]$, with F the Filippov's convexification of f according to (3.8). Existence of the above solution is recently proved by Danca [21]. Same is also stated here.

Theorem 3.3 *Let $\alpha > 0$ and assume f is locally bounded. For such set of initial conditions, there exists $T > 0$ and an absolutely continuous function $x(t)$ on $[0, T]$ such that (3.8) holds for almost $t \in [0, T]$.*

Proof The Approximate Selection Theorem [24] (see J.P. Aubin and A. Cellina) ensures that for any $\varepsilon > 0$, there exists a locally Lipschitzen map f_ε such that

$$\text{Graph}(f_\varepsilon) \subset \text{Graph}(F) + \varepsilon B, \tag{3.9}$$

where B is the unitary ball centered at the origin because the vector field of (3.8) is upper semicontinuous. Thus the standard fractional order differential equation ${}_0D_t^\alpha x(t) = f_\varepsilon(x)$ admits a solution and the proof follows by applying the Convergence Theorem given by Aubin and Cellina [24].

3.3 Sliding Surface Design Methodology for Fractional Order System

State space representation [9] of a fractional order linear time-invariant system is given as

$$\begin{aligned} {}_0D_t^\alpha x(t) &= Ax(t) + B(u + f(t)), \\ y(t) &= Cx(t), \end{aligned} \tag{3.10}$$

where $x(t) \in \mathbb{R}^n, u(t) \in \mathbb{R}^r, f(t) \in \mathbb{R}^r$ and $y(t) \in \mathbb{R}^p$ are states, input, bounded matched disturbances and output vectors of the system and $A \in \mathbb{R}^{n\times n}$, $B \in \mathbb{R}^{n\times r}, C \in \mathbb{R}^{p\times n}$ and α is the fractional commensurate order and the pair (A, B) is controllable.

Lemma 3.4 *There exists a nonsingular transformation matrix T, which defines the state transformation $z(k) = Tx(k)$ such that the system* (3.10) *with matched disturbance f is transformed into the regular form.*

$$ {}_0D_t^\alpha z(t) = TAT^{-1}z(t) + TB(u + f(t)), \tag{3.11}$$

where

$$TAT^{-1} = \begin{bmatrix} A_{11} & A_{12} \\ A_{21} & A_{22} \end{bmatrix}, \quad TB = \begin{bmatrix} 0 \\ B_2 \end{bmatrix}. \tag{3.12}$$

As the pair (A, B) is controllable, the pair (A_{11}, A_{12}) is also controllable.

Using (3.11) and (3.12) one can get

$$\begin{aligned} {}_0D_t^\alpha z_1(t) &= A_{11}z_1(t) + A_{12}z_2(t), \\ {}_0D_t^\alpha z_2(t) &= A_{21}z_1(t) + A_{22}z_2(t) + B_2u(t) + B_2 f, \end{aligned} \tag{3.13}$$

where $z_1 \in \mathbb{R}^{n-m}$, $z_2 \in \mathbb{R}^m$ and $u \in \mathbb{R}^m$. In this paper, the results are applied to the single input case ($m = 1$). However the results can be extended to the multi-input case also. The sliding surface is designed as

$$s(z, t) = C\,{}_0I_t^{1-\alpha}z(t) = {}_0I_t^{1-\alpha}[C_1 \;\; 1]\begin{bmatrix} z_1(t) \\ z_2(t) \end{bmatrix}, \tag{3.14}$$

such that, when the motion is restricted to the manifold, it will be shown that the sliding motion will be governed by the following equation

$$ {}_0D_t^{\alpha} z_1(t) = (A_{11} - A_{12}C_1)z_1(t), \tag{3.15}$$

has the required performance. Using Eq. (3.14) in (3.13) and taking the time derivative of the sliding surface s, the system dynamics (3.13) is transformed to the following form

$$\begin{aligned} {}_0D_t^{\alpha} z_1(t) &= (A_{11} - A_{12}C_1)z_1(t) + A_{12}\, {}_0D_t^{1-\alpha} s, \\ \dot{s} &= C_1\, {}_0D_t^{\alpha} z_1(t) + {}_0D_t^{\alpha} z_2(t) \\ &= C_1 \left\{ (A_{11} - A_{12}C_1)z_1(t) + A_{12}\, {}_0D_t^{1-\alpha} s \right\} \\ &\quad + A_{21}z_1(t) + A_{22}z_2(t) + B_2u(t) + B_2 f. \end{aligned} \tag{3.16}$$

3.3.1 Existence Condition of Sliding Mode

The main aim here, is to design $u(t)$, such that the sliding motion occurs in finite time. For this purpose the control is selected according to the following theorem

Theorem 3.5 *The control input $u(t)$ which is defined as*

$$\begin{aligned} u(t) &= \frac{1}{B_2}\left(v - C_1 \left\{ (A_{11} - A_{12}C_1)z_1(t) + A_{12}\, {}_0D_t^{1-\alpha} s \right\}\right) \\ &\quad - \frac{1}{B_2}(A_{21}z_1(t) + A_{22}z_2(t)) \end{aligned} \tag{3.17}$$

where,

$$v = -k_1 sign(s)) \tag{3.18}$$

with $k_1 > ||B_2 f||$ leads to the establishment of s equal to zero in finite time.

Proof The closed loop system after substituting (3.17) into (3.16)

$$\begin{aligned} {}_0D_t^{\alpha} z_1(t) &= (A_{11} - A_{12}C_1)z_1(t) + A_{12}\, {}_0D_t^{1-\alpha} s, \\ \dot{s} &= v + B_2 f_2 = -k_1 \mathrm{sign}(s) + B_2 f. \end{aligned} \tag{3.19}$$

Now choosing the Lyapunov candidate function as $V = \frac{1}{2}s^2$ and taking the time derivative of V, along the subsystem of (3.19) in s, we get

$$\begin{aligned}\dot{V} &= s\dot{s} = s(-k_1 \text{sign}(s) + B_2 f) \\ &= -k_1|s| + sB_2 f \le -k_1|s| + |s||B_2 f| \\ &\le -(2V)^{\frac{1}{2}}(k_1 - |B_2 f|) \le -\eta(2V)^{\frac{1}{2}}\end{aligned} \tag{3.20}$$

Therefore, using the the designed control (3.17) $s = 0$ in finite time.

3.3.2 Analysis of Sliding Motion

In this subsection the main aim is to find the condition under which the reduced order dynamics during sliding is stable and what actually happens after $s = 0$. For this the following Lemma [17] helps us to derive some conclusion:

Lemma 3.6 *Consider a vector signal $z(t) \in \mathbb{R}^m$. If there exists $T < \infty$ such that*

$${}_0I_t^{1-\alpha} Cz(t) = 0, \ \forall t \ge T \Rightarrow \lim_{t\to\infty} z(t) = 0. \tag{3.21}$$

Proof Applying ${}_0D_t^{1-\alpha}$ to the both side of Eq. (3.14), one can get

$$\begin{aligned}{}_0D_t^{1-\alpha}\,{}_0I_t^{1-\alpha} Cz(t) &= {}_0D_t^{1-\alpha} s, \\ \Rightarrow Cz(t) &= \frac{1}{\Gamma(\alpha)} \frac{d}{dt} \int_0^t \frac{s(\tau)}{(t-\tau)^{1-\alpha}} d\tau.\end{aligned} \tag{3.22}$$

For large values of t, in fact for all $t \ge T$, the above equation reduces to

$$Cz(t) = \frac{1}{\Gamma(\alpha)} \frac{d}{dt} \int_0^T \frac{s(\tau)}{(t-\tau)^{1-\alpha}} d\tau, \tag{3.23}$$

because $s(t)$ clips the upper limit of the integral. The time variable t can now be seen as a parameter, and under mild regularity conditions for the function $s(t)$ (namely, function $p(t, \tau) = s(\tau)(t-\tau)^{\alpha-1}$ must be absolutely integrable in the interval $\tau \in [0, T]$), then the derivative operator can be brought inside the integral, yielding

$$\begin{aligned}Cz(t) &= \frac{1}{\Gamma(\alpha)} \int_0^T s(\tau) \frac{\partial}{\partial t} \frac{1}{(t-\tau)^{1-\alpha}} d\tau \\ &= \frac{\alpha - 1}{\Gamma(\alpha)} \int_0^T s(\tau) \frac{1}{(t-\tau)^{2-\alpha}} d\tau.\end{aligned} \tag{3.24}$$

Now, it is clear that

$$|Cz(t)| \leq \frac{1-\alpha}{\Gamma(\alpha)} \int_0^T |s(\tau)| \frac{1}{|t-\tau|^{2-\alpha}} d\tau. \tag{3.25}$$

Since the right hand side tends to zero when t approaches infinity and also $C \neq 0 > 0$ $\Rightarrow |z(t)|$ must be zero.

Also when t approaches infinity ${}_0D_t^{1-\alpha}s \Rightarrow 0$ in (3.22). Therefore the reduced order dynamics is given by Eq. (3.15).

Lemma 3.7 *Consider the system*

$${}_0D_t^{\alpha} z_1(t) = (A_{11} - A_{12}C_1)z_1(t) = \tilde{A}z_1(t) \tag{3.26}$$

where $\tilde{A} = (A_{11} - A_{12}C_1)$ *and let* $\lambda_i(\tilde{A})(i = 1, 2, .., n-1)$ *be the eigenvalues of matrix* $\tilde{A}$. *The system is asymptotically stable if and only if the following condition holds*

$$|arg(\lambda_i(\tilde{A}))| > \alpha\frac{\pi}{2} \tag{3.27}$$

Proof Proof of Lemma is same as mentioned in [25, 26].

3.3.3 Generalized LMI Based Sliding Surface Design for Fractional Order System

3.3.3.1 D^{α} Stability of Fractional Order System

For the particular value of $\alpha = 1$, D^{α} stability has been given for the integer order system in the following way :

Definition 3.8 A dynamical system represented as $\dot{x} = Ax$, where $x \in \mathbb{R}^n$ and $A \in \mathbb{R}^{n \times n}$ is called D-stable if $\lambda(A)$ (Eigen value of A) lie in D, where D represents the complex left-half plain.

In order to determine D-stability of the integer order systems, the following well established Lyapunov inequality can be applied

$$A^T P + PA < -Q, \quad P = P^T > 0, \quad Q = Q^T \geq 0. \tag{3.28}$$

However, for fractional order system stable region is not the left -half plain, therefore we may need a modified Lyapunov inequality. The Lyapunov inequality for D^{α}-stable system is given in the following way :

Lemma 3.9 [25, 27, 28] *A fractional order system of order* $1 < \alpha < 2$ *is asymptotically* D^{α}*-stable, and the inequality* $|arg(\lambda(A))| > \alpha\frac{\pi}{2}$ *is satisfied if and only if there exists a symmetric positive definite matrix* P *such that*

$$\begin{bmatrix} (A^T P + PA) \, sin(\theta) & (A^T P - PA) \, cos(\theta) \\ (PA - A^T P) \, cos(\theta) & (A^T P + PA) \, sin(\theta) \end{bmatrix} < -Q \tag{3.29}$$

where $Q \geq 0$, $\theta = min\left\{\pi - \alpha\frac{\pi}{2}, \frac{\pi}{2}\right\}$, *or equivalently,*

$$\beta A^T P + \beta^* PA < -Q. \tag{3.30}$$

where $\beta = e^{j(\alpha-1)\frac{\pi}{2}} = cos(1-\alpha)\frac{\pi}{2} + j sin(1-\alpha)\frac{\pi}{2}$, *and* $(.)^*$ *denotes the complex conjugate transpose.*

For the order $0 < \alpha < 1$, *a fractional-order system is asymptotically* D^{α}*-stable if and only if there exists a symmetric positive definite matrix* P_1 *and* P_2 *such that*

$$\beta A^T P_1 + \beta^* P_1 A + \beta A^T P_2 + \beta^* P_2 A < -Q. \tag{3.31}$$

Based on the Lemma 3.9 the following sliding surface

$$s(z, t) = C \, {}_0I_t^{1-\alpha} z(t) = {}_0I_t^{1-\alpha} [C_1 \;\; 1] \begin{bmatrix} z_1(t) \\ z_2(t) \end{bmatrix}, \tag{3.32}$$

is designed, such that, when the motion is restricted to the manifold, the reduced-order model

$${}_0D_t^{\alpha} z_1(t) = (A_{11} - A_{12}C_1) z_1(t), \tag{3.33}$$

has the required performance. For the above specified purpose C_1 is given by the following Theorem :

Theorem 3.10 C_1 *is selected as* $C_1 = \frac{1}{2} A_{12}^T P$, *where positive definite symmetric matrix* P *satisfy the following inequality*

$$2Re\{\beta\} A_{11}^T P + 2Re\{\beta\} PA_{11} - A_{12} A_{12}^T < -Q, \tag{3.34}$$

where Q *is also positive definite symmetric matrix. If any such* P *exists then, closed loop system* (3.33) *is stable.*

Proof The system (3.10) is controllable which implies the pair (A_{11}, A_{12}) is also controllable. The controllability of the pair (A_{11}, A_{12}) guarantees that the pair $(-\rho I - A_{11}, A_{12})$ is also controllable for every $\rho \in R$ (see rank condition for controllability of fractional order systems in [29]). Also one can note that A_{11}^T and $-\rho I - A_{11}^T$ have exactly the same eigenvectors.

$$A_{11}^{\top}x = \lambda x \Leftrightarrow \left(-\rho I - A_{11}^{\top}\right) x = (-\rho - \lambda)\, x, \tag{3.35}$$

where λ represents the eigenvalue of $A_{11}^{\top}$, and let $x \neq 0$ be the corresponding eigenvector. Also, one can choose $\rho > 0$ sufficiently large, so that $\left(-\rho I - A_{11}^{\top}\right)$ is a D^{α} stable matrix. From Lemma 3.9, we can assume that there exists a symmetric positive definite matrix P such that

$$\begin{aligned} \beta P A_{11}^{\top} + \beta^* A_{11} P + \beta^* P A_{11}^{\top} + \beta A_{11} P &< -Q, \\ \Rightarrow 2\mathrm{Re}\{\beta\} A_{11} P + 2\mathrm{Re}\{\beta\} P A_{11}^{\top} &< -Q. \end{aligned} \tag{3.36}$$

The matrix A_{11} is asymptotically D^{α}-stable. Let us assume that there exists a positive-definite matrix Q_1 such that

$$\begin{aligned} \left(-\rho I - 2\mathrm{Re}\{\beta\} A_{11}\right) Q_1 + Q_1 \left(-\rho I - 2\mathrm{Re}\{\beta\} A_{11}\right)^{\top} &= -A_{12} A_{12}^{\top}, \\ \Rightarrow 2\mathrm{Re}\{\beta\} A_{11} Q_1 + 2\mathrm{Re}\{\beta\} Q_1 A_{11}^{\top} - A_{12} A_{12}^{\top} &= -2\rho Q_1. \end{aligned} \tag{3.37}$$

Pre and post multiplying on both sides by $P := Q_1^{-1} > 0$, one obtains

$$\begin{aligned} 2\mathrm{Re}\{\beta\} P A_{11} + 2\mathrm{Re}\{\beta\} A_{11}^{\top} P - P A_{12} A_{12}^{\top} P &= -2\rho P, \\ \Rightarrow 2\mathrm{Re}\{\beta\} P (A_{11} - A_{12} C_1) + 2\mathrm{Re}\{\beta\} (A_{11} - A_{12} C_1)^{\top} P &= -2\rho P, \end{aligned} \tag{3.38}$$

where $C_1 := \frac{1}{2} A_{12}^{\top} P$. Since $P > 0$, $2\rho P > 0$, and from Lemma 3.9, $(A_{11} - A_{12} C_1)$ is a D^{α}-stability matrix. This completes the proof.

3.4 Fraction Order Differential Equation with Discontinuous Right-Hand Side

Consider the following fractional order differential equation

$${}_0D_t^{\alpha} s = -k\mathrm{sign}(s) + d, \tag{3.39}$$

where $s \in \mathbb{R}$, $k > 0$ and $|d| \leq d_0 \in \mathbb{R}$ is a bounded disturbance. For proving the stability of (3.39), choose a Lyapunov candidate function $V = s^2$. According to the Leibniz rule of fractional differentiation, the αth-order time derivative of V can be given as [30]

$${}_0D_t^{\alpha} V = s\, {}_0D_t^{\alpha} s + \Delta, \tag{3.40}$$

where Δ is given as

$$\Delta := \sum_{k=1}^{\infty} \frac{\Gamma(1+\alpha)}{\Gamma(1+k)\Gamma(1-k+\alpha)} {}_0D_t^{\alpha} s \; {}_0D_t^{\alpha-k} s. \tag{3.41}$$

Also, one can easily prove that

$$|\Delta| := \left| \sum_{k=1}^{\infty} \frac{\Gamma(1+\alpha)}{\Gamma(1+k)\Gamma(1-k+\alpha)} {}_0D_t^{\alpha} s \; {}_0D_t^{\alpha-k} s \right| \leq d_1 |s| . \tag{3.42}$$

for some positive d_1. Substituting ${}_0D_t^{\alpha} s$ from (3.39) to (3.40), one can write

$$\begin{aligned} {}_0D_t^{\alpha} V &= s\left(-k\text{sign}(s) + d\right) + \Delta \\ &= -k|s| + sd + \Delta. \end{aligned} \tag{3.43}$$

Using the norm inequality $sd \leq |s||d|$ and substituting the maximum value of Δ from the Eq. (3.42), one can write

$$\begin{aligned} {}_0D_t^{\alpha} V &\leq -k|s| + |s|\,|d| + d_1|s| \\ &\leq -|s|\left(k - |d| - d_1\right). \end{aligned} \tag{3.44}$$

Clearly, if $k > |d| + d_1 \Rightarrow {}_0D_t^{\alpha} V < 0$, and this proves that the system given by (3.39) is Mittag-Leffler stable, which implies that the trajectories in the phase space are attracted by the subspace (manifold) described by $s = 0$.

Now, we have to prove that the first hitting to the switching surface occurs in finite time, which is denoted by T. Applying the fractional integral to both the side of (3.39)

$${}_0I_t^{\alpha} {}_0D_t^{\alpha} s = -{}_0I_t^{\alpha}(k\text{sign}(s)) + {}_0I_t^{\alpha} d. \tag{3.45}$$

Using the following relations

$${}_0I_t^{\alpha} {}_0D_t^{\alpha} s = s(t) - {}_0D_t^{\alpha-1} s(0) \frac{t^{\alpha-1}}{\Gamma(\alpha)}, \tag{3.46}$$

and

$${}_0I_t^{\alpha} c = c \frac{t^{\alpha}}{\Gamma(1+\alpha)}. \tag{3.47}$$

One can write (3.45) after finite time $t = T$ as

$$s(T) - {}_0D_t^{\alpha-1} s(0) \frac{T^{\alpha-1}}{\Gamma(\alpha)} = -k\text{sign}(s(0)) \frac{T^{\alpha}}{\Gamma(1+\alpha)} + {}_0I_t^{\alpha} d. \tag{3.48}$$

Noting that $s(t) = 0$ at $t = T$, multiplying both the sides of (3.48) by $\text{sign}(s(0))$, we have

$$-{}_0D_t^{\alpha-1}s(0)\text{sign}(s(0))\frac{T^{\alpha-1}}{\Gamma(\alpha)} = -k\frac{T^\alpha}{\Gamma(1+\alpha)} + {}_0I_t^\alpha(\text{sign}(s(0))d). \tag{3.49}$$

Using the inequality

$$\begin{aligned} {}_0I_t^\alpha(\text{sign}(s(0))d) &\leq {}_0I_t^\alpha\,|d| \leq {}_0I_t^\alpha d_0 \\ &= d_0\frac{T^\alpha}{\Gamma(1+\alpha)}. \end{aligned} \tag{3.50}$$

Equation (3.49) can be further written as

$$\begin{aligned} -{}_0D_t^{\alpha-1}s(0)\text{sign}(s(0))\frac{T^{\alpha-1}}{\Gamma(\alpha)} &\leq -k\frac{T^\alpha}{\Gamma(1+\alpha)} + d_0\frac{T^\alpha}{\Gamma(1+\alpha)} \\ &\leq -\frac{T^\alpha}{\Gamma(1+\alpha)}(k-d_0). \end{aligned} \tag{3.51}$$

Taking negative sign out from (3.51)

$$\begin{aligned} {}_0D_t^{\alpha-1}s(0)\text{sign}(s(0))\frac{T^{\alpha-1}}{\Gamma(\alpha)} &\geq \frac{T^\alpha}{\Gamma(1+\alpha)}(k-d_0) \\ \Rightarrow T &\leq \frac{\Gamma(1+\alpha){}_0D_t^{\alpha-1}s(0)\text{sign}(s(0))}{\Gamma(\alpha)(k-d_0)}. \end{aligned} \tag{3.52}$$

One can easily check that the above time is always finite.

Remark 3.11 [31] If one views the reaching law using Caputo definition,

$$s\;{}_0D_t^\eta s = \frac{s}{\Gamma(1-\eta)}\int_0^t \frac{Ds(\tau)}{(t-\tau)^\eta}d\tau. \tag{3.53}$$

This relation stipulates that $s > 0$, Ds (the first derivative of s) must be negative to have $s\;{}_0D_t^\eta s < 0$ or alternatively for $s < 0$, Ds must be positive to have $s\;{}_0D_t^\eta s < 0$. Therefore for stability point of view an approximately designed control law via $s\;{}_0D_t^\eta s < 0$ is sufficient for sliding mode of system like (3.39). This explanation of inferring the stability shows that the stability requirement $s\dot{s} < 0$ of the integer order design forces naturally $s\;{}_0D_t^\eta s < 0$. Therefore choosing $\frac{1}{2}s^2$ as Lyapunov function is still valid for both fractional and integer order systems, only requirement is that η must lies between 0 and 1.

3.4.1 Design of Sliding Mode Control Using Fractional Reaching Law Approach

The sliding surface is designed as

$$s(z,t) = [C_1 \;\; 1]\begin{bmatrix} z_1(t) \\ z_2(t) \end{bmatrix}, \tag{3.54}$$

such that, when the motion of the system (3.13) is restricted to the manifold, the reduced-order dynamics is governed by the following equation :

$${}_0D_t^{\alpha} z_1(t) = (A_{11} - A_{12}C_1)z_1(t), \tag{3.55}$$

has the required performance. Using Eq. (3.54), and taking the fractional derivative of the sliding surface s, the system dynamics (3.13) is transformed to the following form

$$\begin{aligned} {}_0D_t^{\alpha} z_1(t) &= (A_{11} - A_{12}C_1)z_1(t) + A_{12}s, \\ {}_0D_t^{\alpha} s &= C_1 \, {}_0D_t^{\alpha} z_1(t) + {}_0D_t^{\alpha} z_2(t) \\ &= C_1 \{(A_{11} - A_{12}C_1)z_1(t) + A_{12}s\} \\ &\quad + A_{21}z_1(t) + A_{22}z_2(t) + B_2u(t) + B_2 f. \end{aligned} \tag{3.56}$$

3.4.2 Existence Condition of Sliding Mode

The main aim here, is to design $u(t)$, such that sliding motion occurs in finite time. For this purpose the control is selected according to the following theorem :

Theorem 3.12 *The control input $u(t)$ which is defined as*

$$\begin{aligned} u(t) = &\frac{1}{B_2}(\nu - C_1 \{(A_{11} - A_{12}C_1)z_1(t) + A_{12}s\}) \\ &- \frac{1}{B_2}(A_{21}z_1(t) + A_{22}z_2(t)), \end{aligned} \tag{3.57}$$

where,

$$\nu = -k_1 sign(s)), \tag{3.58}$$

with $k_1 > ||B_2 f||$ leads to the establishment of s equal to zero in finite time.

Proof The closed loop system after substituting (3.57) into (3.56)

$$\begin{aligned} {}_0D_t^{\alpha} z_1(t) &= (A_{11} - A_{12}C_1)z_1(t) + A_{12}s, \\ {}_0D_t^{\alpha} s &= v + B_2 f_2 = -k_1 \text{sign}(s) + B_2 f. \end{aligned} \tag{3.59}$$

Using the results of Sect. 3.4, $s = 0$ in finite time. Therefore the reduced order dynamics is given by Eq. (3.55). The reduced order system is asymptotically stable if sliding variable C_1 is designed based on the Lemma 3.7.

Remark 3.13 A few important properties of fractional derivatives and integrals which is necessary to develop the further results of this chapter are the following [32]:

- For $\alpha = n$, where n is an integer, the operation $(D_{t_s}^{\alpha} f)(t)$ gives the same result as classical differentiation of integer order n.
- For $\alpha = 0$ the operation $(D_{t_s}^{\alpha} f)(t)$ is the identity operator:

$$(D_{t_s}^{\alpha} f)(t) = f(t).$$

- Fractional differentiation and integration are linear operations:

$$D_{t_s}^{\alpha}(af(t) + bg(t)) = a(D_{t_s}^{\alpha} f)(t) + b(D_{t_s}^{\alpha} g)(t). \tag{3.60}$$

- Invertibility
  The RL fractional derivative of order $\alpha \in (0, 1)$ is the left inverse of the RL fractional integral of the same order

$${}^{RL}D_{t_s}^{\alpha} I_{t_s}^{\alpha} f(t) = f(t). \tag{3.61}$$

  The opposite is however is not true. For Caputo's derivative, the following is true

$$I_{t_s}^{\alpha}\, {}^{C}D_{t_s}^{\alpha} f(t) = f(t) - f(t_s). \tag{3.62}$$

  where, t_s is the initial time.

In the next section a brief review on point to point control of fractional differential linear control systems in the form of $n\alpha$-integrator [33] is presented, which is essential for the analysis of the finite time stabilization of the fractional order uncertain chain of integrator.

3.5 A Brief Review on Point to Point Control

Consider a control system of order $n\alpha$, given by

$$(D_{t_s}^{n\alpha} z)(t) = v(t), \tag{3.63}$$

where $\alpha \in \mathbb{R}$, $0 < \alpha < 1$, and $n \in \mathbb{N}_+$, such that $n\alpha < 1$ with initial conditions

$$(I_{t_s}^{1-\alpha} D_{t_s}^{k\alpha} z)(t_s) = w_k, \quad w_k \in \mathbb{R}, \quad k = 0, \ldots, n-1 \tag{3.64}$$

where $z(t)$ is a scalar solution, $v(t)$ is a scalar control function, and $D_{t_s}^{k\alpha}$ satisfies the following

$$D_{t_s}^{k\alpha} z = D_{t_s}^{\alpha} D_{t_s}^{(k-1)\alpha} z, \quad k = 2, 3, \ldots, n. \tag{3.65}$$

Initial conditions (3.64) are equivalent to [34]:

$$\lim_{t \to t_s} (t - t_s)^{1-\alpha} (D_{t_s}^{k\alpha} z)(t) = \frac{w_k}{\Gamma(\alpha)}, \quad w_k \in \mathbb{R}, \quad k = 0, \ldots, n-1. \tag{3.66}$$

The aim of the control is to steer system (3.63) from the starting point

$$\begin{aligned} Z(t_s) &:= \left(z(t_s), (D_{t_s}^{\alpha} z)(t_s), \ldots, (D_{t_s}^{(n-1)\alpha} z)(t_s) \right)^T \\ &= (z_{s0}, z_{s1}, \ldots, z_{sn-1})^T := Z_s, \end{aligned} \tag{3.67}$$

to the final point

$$\begin{aligned} Z(t_f) &:= \left(z(t_f), (D_{t_s}^{\alpha} z)(t_f), \ldots, (D_{t_s}^{(n-1)\alpha} z)(t_f) \right)^T \\ &= (z_{f0}, z_{f1}, \ldots, z_{fn-1})^T := Z_f, \end{aligned} \tag{3.68}$$

in a finite time interval $t_f - t_s$. The initial condition given in (3.64) corresponding to Z_s, can be calculate from (3.66) as

$$\begin{aligned} w_k &= \Gamma(\alpha) \lim_{t \to t_s} (t - t_s)^{1-\alpha} \lim_{t \to t_s} (D_{t_s}^{k\alpha} z)(t) \\ &= \Gamma(\alpha) \lim_{t \to t_s} (t - t_s)^{1-\alpha} (D_{t_s}^{k\alpha} z)(t_s) \\ &= \Gamma(\alpha) \lim_{t \to t_s} (t - t_s)^{1-\alpha} z_{sk} \quad \text{for } k = 0, \ldots, n-1. \end{aligned} \tag{3.69}$$

Lemma 3.14 *A control $v(t)$ that steers the system* (3.63) *from starting point* (3.67) *to final point* (3.68) *is of the form*

$$v(t) = (D_{t_s}^{n\alpha} \varphi)(t), \tag{3.70}$$

where φ is an arbitrary C^n function satisfying

$$(D_{t_s}^{k\alpha} \varphi)(t_s) = z_{sk}, \; (D_{t_s}^{k\alpha} \varphi)(t_f) = z_{fk}, \quad 0 \le k \le n-1, \tag{3.71}$$

i.e.,

$$\begin{aligned}\left(\varphi(t_s), \ldots, (D_{t_s}^{(n-1)\alpha}\varphi)(t_s)\right)^T &= Z_s \\ \left(\varphi(t_f), \ldots, (D_{t_s}^{(n-1)\alpha}\varphi)(t_f)\right)^T &= Z_f.\end{aligned} \tag{3.72}$$

Proof After applying control (3.70) on the system (3.63),

$$(D_{t_s}^{n\alpha}z)(t) = (D_{t_s}^{n\alpha}\varphi)(t). \tag{3.73}$$

Integrating both sides of (3.73) by means of $I_{t_s}^{\alpha}$,

$$(I_{t_s}^{\alpha}D_{t_s}^{\alpha}D_{t_s}^{(n-1)\alpha}z)(t) = (I_{t_s}^{\alpha}D_{t_s}^{\alpha}D_{t_s}^{(n-1)\alpha}\varphi)(t). \tag{3.74}$$

We get,

$$\begin{aligned}&(D_{t_s}^{(n-1)\alpha}z)(t) - \frac{(I_{t_s}^{1-\alpha}D_{t_s}^{(n-1)\alpha}z)(t_s)}{\Gamma(a)}(t-t_s)^{(1-\alpha)} \\ &= (D_{t_s}^{(n-1)\alpha}\varphi)(t) - \frac{(I_{t_s}^{1-\alpha}D_{t_s}^{(n-1)\alpha}\varphi)(t_s)}{\Gamma(a)}(t-t_s)^{(1-\alpha)}.\end{aligned} \tag{3.75}$$

Since, system (3.63) starts from (3.67), and (3.71) holds, and considering (3.66) we get

$$(D_{t_s}^{(n-1)\alpha}z)(t) = (D_{t_s}^{(n-1)\alpha}\varphi)(t). \tag{3.76}$$

Analogously, consecutive integrations of (3.76) by means of $I_{t_s}^{\alpha}$, yields (for all n integrations)

$$(D_{t_s}^{k\alpha}z)(t) = (D_{t_s}^{k\alpha}\varphi)(t), \quad k = 0, \ldots, n-1, \tag{3.77}$$

and then

$$(D_{t_s}^{k\alpha}z)(t_f) = (D_{t_s}^{k\alpha}\varphi)(t_f) = z_{fk}, \quad k = 0, \ldots, n-1. \tag{3.78}$$

A possible choice of function $\varphi(t)$ is

$$\varphi(t) = \sum_{i=0}^{2n-1} a_i \left(I_{t_s}^{i\alpha}1\right)(t), \tag{3.79}$$

where $\left(I_{t_s}^{i\alpha}1\right)(t)$ is given by

$$\left(I_{t_s}^{i\alpha}1\right)(t) = \frac{1}{\Gamma(i\alpha+1)}(t-t_s)^{i\alpha}, \quad 0 \le i \le 2n-1$$
$$\left(I_{t_s}^{0}1\right)(t) = 1, \tag{3.80}$$

and satisfying (3.71). Since, for a function of type $(t-t_s)^{i\alpha}$ is

$$\left(D_{t_s}^{n\alpha}(t-t_s)^{i\alpha}\right)(t) = \left(\underbrace{D_{t_s}^{\alpha}\dots D_{t_s}^{\alpha}}_{n-\text{times}}(t-t_s)^{i\alpha}\right)(t)$$
$$\text{for } i\alpha + 1 > 0, \tag{3.81}$$

which is always satisfied as we have $i = 0, \dots, 2n-1$, $n\alpha < 1$. Hence we have,

$$(D_{t_s}^{n\alpha} I_{t_s}^{i\alpha} 1)(t) = (I_{t_s}^{(i-n)\alpha} 1)(t). \tag{3.82}$$

Therefore,

$$v(t) = \sum_{i=0}^{2n-1} a_i \left(I_{t_s}^{(i-n)\alpha} 1\right)(t). \tag{3.83}$$

Remark 3.15 The open loop control which is discussed in [33], is not able to stabilize the uncertain fractional chain of integrator. Therefore in the next section a novel feedback controller based on fractional order sliding mode is proposed [35].

3.6 Stabilization of Uncertain Fractional Chain of Integrator

Consider a fractional-order system of the form:

$$\begin{aligned}
(D_{t_s}^{\alpha} x_1)(t) &= x_2(t) \\
(D_{t_s}^{\alpha} x_2)(t) &= x_3(t) \\
&\vdots \\
(D_{t_s}^{\alpha} x_{n-1})(t) &= x_n(t) \\
(D_{t_s}^{\alpha} x_n)(t) &= u + \phi.
\end{aligned} \tag{3.84}$$

where $x(t) \in \mathbb{R}^n$ represents the state vector, $u(t) \in \mathbb{R}$; ϕ is a bounded uncertainty and α is positive integer and also $n\alpha - 1 < 1$. The derivative of ϕ is also assumed to be bounded. Our task is to design a finite time stable controller for above system.

System (3.84) can be represented in the following form

$$\begin{aligned}(D_{t_s}^{(n-1)\alpha} z_1)(t) &= z_2,\\ (D_{t_s}^{\alpha} z_2)(t) &= u + \phi,\end{aligned} \tag{3.85}$$

where $z_1 = x_1(t)$, $z_2 = x_n(t)$. Following results are used for the above transformation.

Remark 3.16 • Miller and Ross [36]

$$\begin{aligned}(D_{t_s}^{\alpha} f)(t) &= D_{t_s}^{\alpha_1} D_{t_s}^{\alpha_2} \cdots D_{t_s}^{\alpha_n} f(t),\\ \alpha &= \alpha_1 + \alpha_2 + \cdots + \alpha_n,\\ \alpha_i &< 1, i = 1, 2, \ldots, n\end{aligned} \tag{3.86}$$

• Composition Property given by Miller and Ross [36]

$$\begin{aligned}(D_{t_s}^{\alpha} f)(t) &= D_{t_s}^{m} D_{t_s}^{\alpha-m} f(t),\\ m &= 1, 2, \ldots, n\end{aligned} \tag{3.87}$$

Theorem 3.17 *Consider the fractional order system* (3.85). *Let n be the total number of fractional integrator and t_f ($t_s < t_f < \infty$) the desired convergence time. Let $S \in \mathbb{R}$ a function defined as*

$$S(z_2, t) = \begin{cases} z_2 - v(t) & \text{for } t_s \le t \le t_f\\ z_2 & \text{for } t > t_f\end{cases} \tag{3.88}$$

with $v(t)$ defined by (3.83). *Then, the control input u is*

$$u = -I_{t_s}^{1-\alpha}(K sign(S(z_2, t))), \tag{3.89}$$

with

$$K \ge |D_{t_s}^{(1-\alpha)} \phi| + |\dot{v}(t)|, K > 0, \tag{3.90}$$

leads to the establishment of finite time stabilization of uncertain chain of fractional order integrators.

Proof Substituting (3.88) in (3.85) for the time interval $t_s \le t_f$, one gets

$$(D_{t_s}^{(n-1)\alpha} z_1)(t) = S(z_2, t) + v(t). \tag{3.91}$$

Now differentiating $S(z_2, t)$

$$\begin{aligned}\dot{S}(z_2, t) &= \dot{z_2} - \dot{v}(t)\\ &= D_{t_s}^{1-\alpha}(D_{t_s}^{\alpha}(z_2)) - \dot{v}(t)\\ &= D_{t_s}^{1-\alpha}(u + \phi) - \dot{v}(t)\\ &= D_{t_s}^{1-\alpha}u + (D_{t_s}^{1-\alpha}\phi - \dot{v}(t)).\end{aligned} \tag{3.92}$$

By substituting the value of u from (3.89), we get

$$\dot{S}(z_2, t) = -K\text{sgn}(S(z_2, t)) + (D_{t_s}^{1-\alpha}\phi - \dot{v}(t)). \tag{3.93}$$

Select the gain K, so that the η-reachability condition of sliding mode is satisfied

$$\begin{aligned}S(z_2, t)\dot{S}(z_2, t) = S(z_2, t)(-K\text{sgn}(S(z_2, t)\\ + (D_{t_s}^{1-\alpha}\phi - \dot{v}(t))) \leq -\eta\,|S(z_2, t)|\,,\end{aligned} \tag{3.94}$$

where $\eta > 0$ is a positive real number. After achieving the sliding mode $S(z_2, t) = 0$, reduced order dynamics from Eq. (3.91) is given as:

$$(D_{t_s}^{(n-1)\alpha}z_1)(t) = v(t), \tag{3.95}$$

The control law can now be designed according to the available result (3.70). First $(n-1)$ states are stabilized but the last state z_2 is yet to be stabilized. Equation (3.88), implies that after $t > t_f$, $S(z_2, t) = z_2$. Hence from (3.89) control input u is switched to $u = -I_{t_s}^{1-\alpha}(K\text{sign}(z_2))$ and the dynamics of z_2 is given as

$$(D_{t_s}^{\alpha}z_2)(t) = -I_{t_s}^{1-\alpha}(K\text{sign}(z_2)) + \phi. \tag{3.96}$$

Applying $D_{t_s}^{1-\alpha}$ on both sides of (3.96), one can get

$$\dot{z_2} = -K\text{sign}(z_2) + D_{t_s}^{1-\alpha}\phi. \tag{3.97}$$

The above equation is finite time stable because $K \geq |D_{t_s}^{(1-\alpha)}\phi|$. Hence, the motive of finite time stabilization of uncertain chain of integrators is achieved.

This completes the proof.

3.7 Numerical Example

Example Consider the following system

$$D_{t_s}^{\frac{1}{4}} x_1 = x_2$$
$$D_{t_s}^{\frac{1}{4}} x_2 = x_3 \qquad (3.98)$$
$$D_{t_s}^{\frac{1}{4}} x_3 = u + \phi.$$

Our aim is to calculate a switching control for the above system so that the system starts at $t_s = 1$ with initial point $x(2, -3, 5)$ and moves to the final point (0, 0, 0) at time $\mathrm{t}_f = 5$. Desired sliding surface for the above goal is

$$S(x_3, t) = \begin{cases} x_3 - v(t) & \text{for } 1 \leq t \leq 5 \\ x_3 & \text{for } t > 5 \end{cases} \qquad (3.99)$$

where $v(t)$ is define latter. Then, the control input u which is given as

$$u = -I_{t_s}^{1-\alpha}(K\operatorname{sign}(S(x_3, t))), \qquad (3.100)$$

is able to stabilize the system (3.98) in finite time. Same as in Eq. (3.95) after applying the control (3.100), $S(z_2, t) = 0$ and $x_3 = v(t)$ for the time $t_s \leq t \leq t_f$ and the system (3.98) is reduced to

$$D_{t_s}^{\frac{1}{4}} x_1 = x_2$$
$$D_{t_s}^{\frac{1}{4}} x_2 = v(t). \qquad (3.101)$$

Denoting $z = x_1$ (and as a consequence $D_{t_s}^{\frac{1}{4}} z = x_2$), we get

$$D_{t_s}^{\frac{1}{4}} D_{t_s}^{\frac{1}{4}} z = v(t). \qquad (3.102)$$

Control for finite time stabilization of (3.101) is of the form (3.83)

$$v(t) = (D_{t_s}^{\frac{1}{2}} \varphi)(t), \qquad (3.103)$$

$\varphi(t)$, from Eqs. (3.79) and (3.80), is given as

$$\varphi(t) = \sum_{i=0}^{3} a_i \frac{1}{\Gamma(\frac{i}{4} + 1)} (t - 1)^{\frac{i}{4}}, \qquad (3.104)$$

The coefficient a_i, $0 \leq i \leq 3$, are such that

$$\varphi(1) = a_0 = 2$$
$$(D_{t_s}^{\frac{1}{4}} \varphi(1)) = a_1 = -3$$

$$\varphi(5) = \sum_{i=0}^{3} a_i \frac{(\sqrt{2})^i}{\Gamma(\frac{i}{4}+1)} = 0$$
$$(D_{t_s}^{\frac{1}{4}}\varphi(5)) = \sum_{i=0}^{3} a_i \frac{(\sqrt{2})^{(i-1)}}{\Gamma(\frac{1}{4}(i-1)+1)} = 0. \tag{3.105}$$

Solving the Eq. (3.105) $a_0 = 2; a_1 = -3; a_2 = 1.2666; a_3 = -0.0577.$

According to (3.83) ,we have

$$v(t) = (D_{t_s}^{\frac{1}{2}}\varphi)(t) = \sum_{i=0}^{3} a_i \frac{1}{\Gamma(\frac{(i-2)}{4}+1)}(t-1)^{\frac{i-2}{4}}. \tag{3.106}$$

Using $v(t)$, x_1 and x_2 are stabilized in finite time but still x_3 is not stabilized. After $t > 5$, $S(x_3, t) = x_3$, the control input u is switched to $u = -I_{t_s}^{1-\alpha}(K\text{sign}(x_3))$ and x_3 is stabilized in finite time as discussed in the proof of Theorem 3.17.

3.8 Summary

In this chapter different methodologies of fractional order sliding mode control is discussed. The finite time stabilization of a fractional order chain of integrator has been proposed for the first time. The finite time stabilization is achieved by using switched sliding surfaces. The main contribution of this chapter is the design of finite time stabilizing robust controller, which takes care of the uncertainty. The proposed controller can reject matched disturbances. However, this is limited to the class of the systems which have $n\alpha - 1$ lies between 0 and 1.

References

1. Machado, J.T., Kiryakova, V., Mainardi, F.: Recent history of fractional calculus. Commun. Nonlinear Sci. Numer. Simul. **16**(3), 1140–1153 (2011)
2. Li, C., Zhang, F.: A survey on the stability of fractional differential equations. Eur. Phys. J. Spec. Top. **193**(1), 27–47 (2011)
3. Ortigueira, M.D.: An introduction to the fractional continuous-time linear systems: the 21st century systems. IEEE Circuits Syst. Mag. **8**(3), 19–26 (2008)
4. Xue, D., Chen, Y.: A comparative introduction of four fractional order controllers. In: Intelligent Control and Automation, 2002. Proceedings of the 4th World Congress on, vol. 4, pp. 3228–3235 (2002)
5. Agrawal, O.P.: A general formulation and solution scheme for fractional optimal control problems. Nonlinear Dyn. **38**(1–4), 323–337 (2004)
6. Oustaloup, A., Mathieu, B., Lanusse, P.: The crone control of resonant plants: application to a flexible transmission. Eur. J. Control **1**(2), 113–121 (1995)

7. Podlubny, I.: Fractional-order systems and pi/sup/spl lambda//d/sup/spl mu//-controllers. IEEE Trans. Autom. Control **44**(1), 208–214 (1999)
8. Bouafoura, M.K., Braiek, N.B.: Piλdμ controller design for integer and fractional plants using piecewise orthogonal functions. Commun. Nonlinear Sci. Numer. Simul. **15**(5), 1267–1278 (2010)
9. Raynaud, H.F., Zergaınoh, A.: State-space representation for fractional order controllers. Automatica **36**(7), 1017–1021 (2000)
10. Utkin, V., Guldner, J., Shi, J.: Sliding mode control in electro-mechanical systems. CRC Press, Boca Raton (2009)
11. Edwards, C., Spurgeon, S.K.: Sliding mode control: theory and applications. CRC Press, Boca Raton (1998)
12. Utkin, V.I.: Sliding modes in control and optimization, vol. 116. Springer, Berlin (1992)
13. Fridman, L., Levant, A.: Higher order sliding modes. Sliding Mode Control Eng. **11**, 53–102 (2002)
14. Si-Ammour, A., Djennoune, S., Bettayeb, M.: A sliding mode control for linear fractional systems with input and state delays. Commun. Nonlinear Sci. Numer. Simul. **14**(5), 2310–2318 (2009)
15. Efe, M.Ö.: Fractional order sliding mode controller design for fractional order dynamic systems. In: New Trends in Nanotechnology and Fractional Calculus Applications, pp. 463–470. Springer (2010)
16. Ramírez, H.S., Battle, V.F.: A generalized PI sliding mode and PWM control of switched fractional systems. In: Bartolini, G. et al. (eds.) Modern Sliding Mode Control Theory, LNCIS, vol. 375, pp. 201–221. Springer (2008)
17. Pisano, A., Rapaić, M., Usai, E.: Second-order sliding mode approaches to control and estimation for fractional order dynamics. In: Fridman, L. et al. (eds.) Sliding Modes after the First Decade of the 21st Century, LNCIS, vol. 412, pp. 169–197. Springer (2012)
18. Cortes, J.: Discontinuous dynamical systems. IEEE Control Syst. Mag. **28**(3), 36–73 (2008)
19. Filippov, A.F.: Differential equations with discontinuous right-hand side. Matematicheskii sbornik **93**(1), 99–128 (1960)
20. Dieci, L., Lopez, L.: Sliding motion in filippov differential systems: theoretical results and a computational approach. Numer. Anal. **47**(3), 2023–2051 (2009)
21. Danca, M.F.: Approach of a class of discontinuous dynamical systems of fractional order: Existence of solutions. Int. J. Bifurcat. Chaos **21**(11), 3273–3276 (2011)
22. Cernea, A.: Continuous version of filippov's theorem for fractional differential inclusions. Nonlinear Anal. Theor Methods Appl. **72**(1), 204–208 (2010)
23. Danca, M.F.: Numerical approximation of a class of discontinuous systems of fractional order. Nonlinear Dyn. **66**(1–2), 133–139 (2011)
24. Aubin, J.P., Cellina, A.: Differential inclusions: Set-valued maps and viability theory. Springer, New York (1984)
25. Sabatier, J., Moze, M., Farges, C.: Lmi stability conditions for fractional order systems. Comput. Math. Appl. **59**(5), 1594–1609 (2010)
26. Petráš, I.: Stability of fractional-order systems with rational orders: a survey. Fractional Calc. Appl. Anal. **12**(3), 269–298 (2009)
27. Ahn, H.S., Chen, Y.: Necessary and sufficient stability condition of fractional-order interval linear systems. Automatica **44**(11), 2985–2988 (2008)
28. Gutman, S., Jury, E.: A general theory for matrix root-clustering in subregions of the complex plane. IEEE Trans. Autom. Control **26**(4), 853–863 (1981)
29. Matignon, D., D'Andrea-Novel, B.: Some results on controllability and observability of finite-dimensional fractional differential systems. In: Computational Engineering in Systems Applications, vol. 2, pp. 952–956. Citeseer (1996)
30. Senejohnny, D.M., Delavari, H.: Active sliding observer scheme based fractional chaos synchronization. Commun. Nonlinear Sci. Numer. Simul. **17**(11), 4373–4383 (2012)
31. Efe, M.Ö.: Fractional order systems in industrial automation-a survey. IEEE Trans. Industr. Inf. **7**(4), 582–591 (2011)

32. Chen, Y.Q., Petras, I., Xue, D.: Fractional order control-a tutorial. In: American Control Conference, 2009. ACC'09, pp. 1397–1411 (2009)
33. Andrzej, D., Wiktor, M.: Point to point control of fractional differential linear control systems. Advances in Difference Equations 2011
34. Anatoly, A.K., Srivastava, H.M., Trujillo, J.J.: Theory and Applications of Fractional Differential Equations. Elsevier, Amsterdam (2006)
35. Kamal, S., Raman, A., Bandyopadhyay, B.: Finite time stabilization of fractional order uncertain chain of integrator: a sliding mode approach. In: Industrial Technology (ICIT), 2012 IEEE International Conference on, pp. 1132–1135 (2012)
36. Miller, K.S., Ross, B.: An Introduction to The Fractional Calculus and Fractional Differential Equations. John Wiley & Sons, INC (1993)

Chapter 4
Finite Time Stabilization of Fractional Order Systems

4.1 Introduction

The sliding mode control [1, 2] methodology is one such robust control technique which has its roots in the relay control. One of the most intriguing aspects of sliding mode is the discontinuous nature of the control action whose primary function is to switch between two distinctively different structures about some predefined manifold such that a new type of system motion called sliding mode exists in a manifold. This peculiar system characteristic is claimed to result in a superb system performance which includes insensitivity to parameter variations and complete rejection of certain class of disturbances. Furthermore, the system possess new properties which are not present in original system.

Sliding mode contains two phases (a) reaching phase in which the system states are driven from any initial state to reach the switching manifolds (the anticipated sliding modes) in finite time and (b) sliding phase in which the system is induced into the sliding motion on the switching manifolds, i.e., the switching manifolds become attractors. The robustness and order reduction property of sliding mode control comes into picture only after the occurrence of sliding mode. During the reaching phase, however, there is no guarantee [1]. Integral sliding mode [3, 4] though is a variant. It eliminates the reaching phase by enforcing the sliding mode throughout the entire system response.

In the field of fractional order systems, the application of SMC was reported only recently. In these works the philosophy of integer order SMC was extended to fractional order systems [5, 6]. The SMC design for fractional order systems with input and state delays was also proposed [7]. The application of SMC to fractional order system has been reported in [8–10]. In the above cited references, only asymptotic stabilization of states could be achieved. The most recent attempt at combining SMC with fractional order system was done in [11] in which Super-twisting control was used. But there also, only asymptotic stabilization could be achieved.

However, in numerous applications, it is required that the control objective is achieved in finite time. In the case of finite time stabilization, as the name suggests

B. Bandyopadhyay and S. Kamal, *Stabilization and Control of Fractional Order Systems: A Sliding Mode Approach*, Lecture Notes in Electrical Engineering 317,
DOI 10.1007/978-3-319-08621-7_4

the system states reach the system equilibrium in a finite time [12]. It is required for achieving faster convergence, better robustness as well as disturbance rejection properties. It also gives the optimality in settling time of the controlled system [13, 14]. In primary development, finite time convergence property was considered to be only for non-smooth or non-Lipschitz continuous autonomous system [14]. Recently, finite time stability has been further extended to non-autonomous systems [15], switched systems [16], time-delay systems [17, 18], impulsive dynamical systems [19] and stochastic nonlinear systems [20]. In the case of fractional order systems [21], discusses point to point control of fractional order systems in which any initial state is transferred to any final state in a specified finite amount of time by using an open loop control. However, being an open loop control it is limited to a disturbance free environment. Recently a sliding mode control for uncertain chain of fractional integrators is reported [22]. But much more work needs to be done in this area. Our current work proposes a novel and simple solution to the addressed problem.

In this chapter we propose a novel methodology for robust finite time stabilization of fractional order systems using integral sliding mode algorithm. An improved strategy with more general kind of uncertainty is also proposed. Since, sliding mode control is used, the proposed controller makes the system robust to matched uncertainties.

The chapter has been organized in the following way: In Sect. 4.2, some useful properties of fractional calculus have been presented. Section 4.3 reviews the concept of finite-time steering of a controllable fractional linear system from initial to final states in desired time. The main result on finite time stabilization of uncertain chain of integrator with numerical examples is presented in Sects. 4.4 and 4.5. In Sect. 4.5, a more general uncertain fractional order system is considered. Finally, some concluding remarks are included in Sect. 4.6.

4.2 Some Useful Properties of Fractional Order Calculus

A few important properties of fractional derivatives and integrals [23]:

- For $\alpha = m$, where m is an integer, the operation $(D_{t_s}^{\alpha} f)(t)$ gives the same result as classical differentiation of integer order m.
- For $\alpha = 0$ the operation $(D_{t_s}^{\alpha} f)(t)$ is the identity operator:

$$(D_{t_s}^{\alpha} f)(t) = f(t).$$

- Fractional differentiation and integration are linear operations:

$$D_{t_s}^{\alpha}(af(t) + bg(t)) = a(D_{t_s}^{\alpha} f)(t) + b(D_{t_s}^{\alpha} g)(t).$$

- Miller and Ross [24]

$$(D_{t_s}^{\alpha} f)(t) = D_{t_s}^{\alpha_1} D_{t_s}^{\alpha_2}, \ldots, D_{t_s}^{\alpha_n} f(t)$$
$$\alpha = \alpha_1 + \alpha_2 + \cdots + \alpha_n, \;\; \alpha_i < 1, \; i = 1, 2, \ldots, n. \tag{4.1}$$

- Composition property given by Miller and Ross [24]

$$(D_{t_s}^{\alpha} f)(t) = D_{t_s}^{m} D_{t_s}^{\alpha-m} f(t), \; m = 1, 2, \ldots, n. \tag{4.2}$$

- Invertibility: The RL fractional derivative of order $\alpha \in (0, 1)$ is the left inverse of the RL fractional integral of the same order

$$^{RL}D_{t_s}^{\alpha} I_{t_s}^{\alpha} f(t) = f(t).$$

RL definition is considered henceforth unless otherwise specified.

4.3 Point to Point Control of Fractional Differential Systems in the Form of $n\alpha$-Integrator

Consider a control system of order $n\alpha$, given by

$$(D_{t_s}^{n\alpha} z)(t) = v(t), \tag{4.3}$$

where $\alpha \in \mathbb{R}, 0 < \alpha < 1$, and $n \in \mathbb{N}_+$, such that $n\alpha < 1$ with initial conditions

$$(I_{t_s}^{1-\alpha} D_{t_s}^{k\alpha} z)(t_s) = w_k, \quad w_k \in \mathbb{R}, \; k = 0, \ldots, n-1 \tag{4.4}$$

where $z(t)$ is a scalar solution, $v(t)$ is a scalar control function, and $D_{t_s}^{k\alpha}$ satisfies the following

$$D_{t_s}^{k\alpha} z = D_{t_s}^{\alpha} D_{t_s}^{(k-1)\alpha} z, \; k = 2, 3, \ldots, n. \tag{4.5}$$

Initial conditions (4.4) are equivalent to [25]:

$$\lim_{t \to t_s} (t - t_s)^{1-\alpha} (D_{t_s}^{k\alpha} z)(t) = \frac{w_k}{\Gamma(\alpha)}, \quad w_k \in \mathbb{R}, \; k = 0, \ldots, n-1. \tag{4.6}$$

The aim of the control is to steer system (4.3) from the starting point

$$Z(t_s) := \left(z(t_s), (D_{t_s}^{\alpha} z)(t_s), \ldots, (D_{t_s}^{(n-1)\alpha} z)(t_s)\right)^T$$
$$= (z_{s0}, z_{s1}, \ldots, z_{sn-1})^T := Z_s, \tag{4.7}$$

to the final point

$$\begin{aligned} Z(t_f) &:= \left(z(t_f), (D_{t_s}^{\alpha} z)(t_f), \ldots, (D_{t_s}^{(n-1)\alpha} z)(t_f)\right)^T \\ &= (z_{f0}, z_{f1}, \ldots, z_{fn-1})^T := Z_f, \end{aligned} \tag{4.8}$$

in a finite time interval $t_f - t_s$. The initial condition given in (4.4) corresponding to Z_s, can be calculated from (4.6) as

$$\begin{aligned} w_k &= \Gamma(\alpha) \lim_{t \to t_s} (t - t_s)^{1-\alpha} \lim_{t \to t_s} (D_{t_s}^{k\alpha} z)(t) \\ &= \Gamma(\alpha) \lim_{t \to t_s} (t - t_s)^{1-\alpha} (D_{t_s}^{k\alpha} z)(t_s) \\ &= \Gamma(\alpha) \lim_{t \to t_s} (t - t_s)^{1-\alpha} z_{sk} \quad \text{for } k = 0, \ldots, n-1. \end{aligned} \tag{4.9}$$

Lemma 4.1 [21] *A control $v(t)$ that steers the system (4.3) from the starting point (4.7) to the final point (4.8) is of the form*

$$v(t) = (D_{t_s}^{n\alpha} \varphi)(t), \tag{4.10}$$

where φ is an arbitrary C^n function satisfying

$$(D_{t_s}^{k\alpha} \varphi)(t_s) = z_{sk}, \; (D_{t_s}^{k\alpha} \varphi)(t_f) = z_{fk}, \quad 0 \le k \le n-1, \tag{4.11}$$

i.e.,

$$\begin{aligned} \left(\varphi(t_s), \ldots, (D_{t_s}^{(n-1)\alpha} \varphi)(t_s)\right)^T &= Z_s, \\ \left(\varphi(t_f), \ldots, (D_{t_s}^{(n-1)\alpha} \varphi)(t_f)\right)^T &= Z_f. \end{aligned}$$

Proof Detailed proof of the above Lemma is discussed in [21].

There are many possible choice of $\varphi(t)$ as discussed in [21]. One of the possible choices of function $\varphi(t)$ is [21]

$$\varphi(t) = \sum_{i=0}^{2n-1} a_i \left(I_{t_s}^{i\alpha} 1\right)(t), \tag{4.12}$$

where $\left(I_{t_s}^{i\alpha} 1\right)(t)$ is given by

$$\left(I_{t_s}^{i\alpha} 1\right)(t) = \frac{1}{\Gamma(i\alpha + 1)} (t - t_s)^{i\alpha} \quad 0 \le i \le 2n-1, \text{ and } \left(I_{t_s}^{0} 1\right)(t) = 1, \tag{4.13}$$

and satisfying (4.11). Since, for a function of type $(t - t_s)^{i\alpha}$ is

$$\left(D_{t_s}^{n\alpha}(t-t_s)^{i\alpha}\right)(t) = \left(\underbrace{D_{t_s}^{\alpha},\ldots,D_{t_s}^{\alpha}}_{n-\text{times}}(t-t_s)^{i\alpha}\right)(t) \quad \text{for } i\alpha+1>0, \tag{4.14}$$

which is always satisfied as we have $i = 0, \ldots, 2n-1$, $n\alpha < 1$. Hence we have,

$$(D_{t_s}^{n\alpha} I_{t_s}^{i\alpha} 1)(t) = (I_{t_s}^{(i-n)\alpha} 1)(t). \tag{4.15}$$

Therefore,

$$v(t) = \sum_{i=0}^{2n-1} a_i \left(I_{t_s}^{(i-n)\alpha} 1\right)(t), \tag{4.16}$$

Remark 4.2 The open loop control which is discussed in [21], is not able to stabilize the uncertain fractional chain of integrator. Therefore in the next section a novel feedback controller based on fractional order integral sliding mode is proposed.

4.4 Main Result

Consider a fractional-order system of the form:

$$\begin{aligned}
(D_{t_s}^{\alpha} x_1)(t) &= x_2(t) \\
(D_{t_s}^{\alpha} x_2)(t) &= x_3(t) \\
&\vdots \\
(D_{t_s}^{\alpha} x_n)(t) &= u + \rho,
\end{aligned} \tag{4.17}$$

where $x(t) \in \mathbb{R}^n$ represents the state vector, $u(t) \in \mathbb{R}$, ρ is a bounded uncertainty, $\alpha \in \mathbb{R}$, $0 < \alpha < 1$, and $n \in \mathbb{N}_+$, such that $n\alpha < 1$. Our main aim is to design a finite time stable controller for the above uncertain system.
System (4.17) can be represented in the following form

$$(D_{t_s}^{(n-1)\alpha} z_1)(t) = z_2, \quad (D_{t_s}^{\alpha} z_2)(t) = u + \rho, \tag{4.18}$$

where $z_1 = x_1(t)$, $z_2 = x_n(t)$.

Theorem 4.3 *The control input u which is deﬁned as*

$$u = v(t) + u_{in}, \tag{4.19}$$

where, $v(t)$ deﬁned by (4.16), $u_{in} = -K sign(s(z,t))$ with

$$K \geq |\rho|, \quad K > 0, \tag{4.20}$$

and $s \in \mathbb{R}$ be a proposed novel sliding surface as

$$s(z,t) = (I_{t_s}^{1-\alpha} z_2)(t) - (I_{t_s}^{1-\alpha} z_2)(t_s) - \int_{t_s}^{t} (v(x) - (D_{t_s}^{\alpha} z_2)(t_s))dx, \tag{4.21}$$

leads to the establishment of finite time stabilization of uncertain chain of fractional order system (4.18), where n be the total number of fractional integrators and $t_f (t_s < t_f < \infty)$ the desired convergence time.

Proof Time derivative of (4.21) is given as

$$\begin{aligned}\dot{s}(z,t) &= (D_{t_s}^{\alpha} z_2)(t) - (D_{t_s}^{\alpha} z_2)(t_s) - v(t) + (D_{t_s}^{\alpha} z_2)(t_s) \\ &= (D_{t_s}^{\alpha} z_2)(t) - v(t).\end{aligned} \tag{4.22}$$

Choosing the Lyapunov candidate function as $V = \frac{1}{2}s^2(z,t)$ and taking time derivative of V and using (4.22) and (4.18), we get

$$\begin{aligned}\dot{V} &= s(z,t)\dot{s}(z,t) = s(z,t)\left((D_{t_s}^{\alpha} z_2)(t) - v(t)\right) \\ &= s(z,t)\left(u + \rho - v(t)\right).\end{aligned} \tag{4.23}$$

Using the value of u from (4.19), we get

$$\begin{aligned}\dot{V} &= s(z,t)\left(v(t) + u_{in} + \rho - v(t)\right) \\ &= s(z,t)\left(-K\operatorname{sign}(s(z,t)) + \rho\right).\end{aligned} \tag{4.24}$$

The gain K is selected so that the η-reachability condition of sliding mode is satisfied. One gets

$$s(z,t)\dot{s}(z,t) \leq -\eta\,|s(z,t)|\,, \tag{4.25}$$

where $\eta > 0$ is a positive real number. It means that if the inequality (4.20) is satisfied then, the proposed controller conforms the finite time reachability to $s(z,t)$. From Eqs. (4.22) and (4.18),

$$\dot{s}(z,t) = (D_{t_s}^{\alpha} z_2)(t) - v(t) = u + \rho - v(t) = (u_{in} + \rho)\,.$$

When the system is under sliding motion, $s(z,t) = \dot{s}(z,t) = 0$

$$\Rightarrow \dot{s}(z,t) = (u_{in} + \rho) = 0, \quad \Rightarrow u_{in} = -\rho, \tag{4.26}$$

which implies that when the system is under sliding motion, the equivalent value of the switching control (u_{in}) is negative of the disturbance and hence cancels it out.

Now when the system is in sliding mode the overall dynamics of the system (4.18),

$$(D_{t_s}^{(n-1)\alpha} z_1)(t) = z_2$$
$$(D_{t_s}^{\alpha} z_2)(t) = u + \rho = v(t) + u_{in} + \rho = v(t).$$

The disturbance free dynamics of the system becomes

$$(D_{t_s}^{(n-1)\alpha} z_1)(t) = z_2, \quad (D_{t_s}^{\alpha} z_2)(t) = v(t). \tag{4.27}$$

The above equation is similar to (4.3), with n fractional order integrators. An appropriate $v(t)$ can be designed by the known result [21]. Hence, finite time stabilization of $x_1, x_2, \ldots, x_n$ of the uncertain chain of integrator from any initial point to the origin is achieved.

Remark 4.4 The main idea to define integral sliding surface like (4.21) is that as soon as disturbance comes into play, $s(z, t)$ becomes nonzero and at that time discontinuous control term becomes active and rejects the disturbance.

Remark 4.5 For $\alpha = 1$, the switching surface proposed in (4.21) reduces to the classical integral sliding mode control design. Also a fractional differential term is intentionally introduced in the integral part because initial condition of fractional order system is not only dependent on the present state but also dependent on the whole past history. This intentionally introduced term compensates the effect of past history.

4.4.1 Numerical Example

Consider the following system

$$D_{t_s}^{\frac{1}{4}} x_1 = x_2, \quad D_{t_s}^{\frac{1}{4}} x_2 = u + \rho. \tag{4.28}$$

Our aim is to calculate a switching control for the above system so that the system starts at $t_s = 1$ with an initial point $x(2, -3)$ to the final point $(0, 0)$ at time $t_f = 5$. Let the initial value of control be zero. Desired sliding surface for above goal is

$$s(x, t) = (I_{t_s}^{1-\alpha} x_2)(t) - (I_{t_s}^{1-\alpha} x_2)(t_s) - \int_{t_s}^{t} (v(l) - (D_{t_s}^{\alpha} x_2)(t_s))dl, \tag{4.29}$$

where $v(t)$ is define later. Then, the control input u which is given as $u = v(t) - K\text{sign}(s(x, t))$ able to stabilizes the system (4.28) in finite time. Same as (4.26) after applying the u, $s(x, t) = 0$ and equivalent control $u_{in} = (-K\text{sign}(s(x, t)))_{equi} = -\rho$ and the system (4.28) is reduced as

$$D_{t_s}^{\frac{1}{4}} x_1 = x_2, \quad D_{t_s}^{\frac{1}{4}} x_2 = v(t). \tag{4.30}$$

Denoting $z = x_1$ (and as a consequence $D_{t_s}^{\frac{1}{4}} z = x_2$), we get

$$D_{t_s}^{\frac{1}{4}} D_{t_s}^{\frac{1}{4}} z = v(t). \tag{4.31}$$

Control for stabilization of (4.31) is of the form (4.16)

$$v(t) = (D_{t_s}^{\frac{1}{2}} \varphi)(t), \tag{4.32}$$

where $\varphi(t)$, from Eqs. (4.12) and (4.13), is given as

$$\varphi(t) = \sum_{i=0}^{3} a_i \frac{1}{\Gamma(\frac{i}{4}+1)} (t-1)^{\frac{i}{4}}, \tag{4.33}$$

and the coefficient $a_i,\ 0 \le i \le 3$, are such that

$$\begin{aligned} \varphi(1) = a_0 = 2, \quad (D_{t_s}^{\frac{1}{4}} \varphi(1)) &= a_1 = -3, \\ \varphi(5) = \sum_{i=0}^{3} a_i \frac{(\sqrt{2})^i}{\Gamma(\frac{i}{4}+1)} &= 0, \\ (D_{t_s}^{\frac{1}{4}} \varphi(5)) = \sum_{i=0}^{3} a_i \frac{(\sqrt{2})^{(i-1)}}{\Gamma(\frac{1}{4}(i-1)+1)} &= 0. \end{aligned} \tag{4.34}$$

Solving the above equations, we get $a_0 = 2$; $a_1 = -3$; $a_2 = 1.2666$; $a_3 = -0.0577$ According to (4.16), we have

$$v(t) = (D_{t_s}^{\frac{1}{2}} \varphi)(t) = \sum_{i=0}^{3} a_i \frac{1}{\Gamma(\frac{(i-2)}{4}+1)} (t-1)^{\frac{i-2}{4}}. \tag{4.35}$$

Using $v(t)$, x_1 and x_2 are stabilized in finite time as discussed in [21]. Evolution of states is shown in Fig. 4.1 for $K = 3.5$.

Remark 4.6 The result discussed in the above section is only limited to the class of systems where control part is free from the uncertainty. In the next section finite time stabilization of chain of fractional order integrators with more general kind of uncertainty is considered.

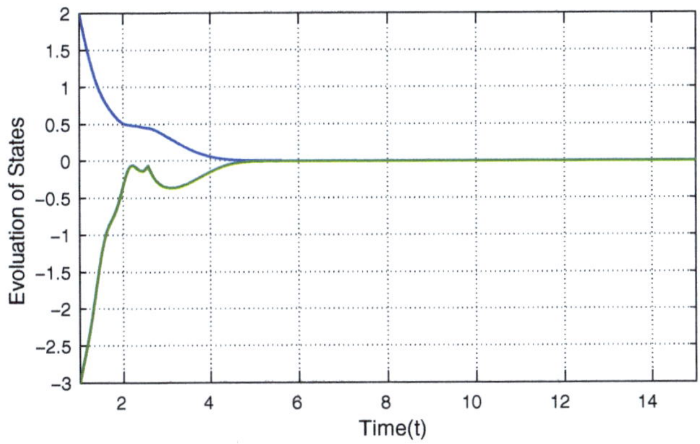

Fig. 4.1 Evolution of states x_1 and x_2

4.5 Finite Time Stabilization of More General Uncertain Fractional Order Systems

Consider a fractional-order system of the form:

$$\begin{aligned}(D^{\alpha}_{t_s}x_1)(t) &= x_2(t)\\ (D^{\alpha}_{t_s}x_2)(t) &= x_3(t)\\ &\vdots\\ (D^{\alpha}_{t_s}x_n)(t) &= \omega + \gamma u(t).\end{aligned} \tag{4.36}$$

where $x(t) \in \mathbb{R}^n$ represents the state vector, $u(t) \in \mathbb{R}$, $\alpha \in \mathbb{R}, 0 < \alpha < 1$, and $n \in \mathbb{N}_+$, such that $n\alpha < 1$. Assume that $\omega = \widehat{\omega} + \Delta\omega$ and $\gamma = \widehat{\gamma} + \Delta\gamma$. Where ω and γ are expressed in terms of the nominal part (as $\widehat{\omega}$ and $\widehat{\gamma}$), and bounded additive uncertainties as $\Delta\omega$ and $\Delta\gamma$. It is also assumed that the uncertain bounded functions satisfy the following inequalities

$$|\Delta\omega - \Delta\gamma\widehat{\gamma}^{-1}\widehat{\omega}| \leq \delta, \quad |\Delta\gamma\widehat{\gamma}^{-1}| \leq 1 - \beta, \tag{4.37}$$

where $\widehat{\gamma}$ is non-singular and δ is a priori known constant and that $0 \leq \beta \leq 1$. Let (4.36), be finite time stabilizable by the control

$$u(t) = \widehat{\gamma}^{-1}\left[-\widehat{\omega} + \widehat{u}(t)\right]. \tag{4.38}$$

After applying the above $u(t)$, Eq. (4.36) can also be written as

$$\begin{aligned}(D^{(n-1)\alpha}_{t_s}z_1)(t) &= z_2,\\ (D^{\alpha}_{t_s}z_2)(t) &= \left(\Delta\omega - \Delta\gamma\widehat{\gamma}^{-1}\widehat{\omega}\right) + \left(1 + \Delta\gamma\widehat{\gamma}^{-1}\right)\widehat{u}(t),\end{aligned} \tag{4.39}$$

where $z_1 = x_1(t), z_2 = x_n(t)$. Our main aim is to design $\widehat{u}(t)$ so that the transformed system (4.39) is finite time stable in spite of any matched uncertainty.

Theorem 4.7 *The control input $\widehat{u}(t)$ which is defined as*

$$\widehat{u}(t) = v(t) + u_{in}, \tag{4.40}$$

where, $u_{in} = -K sign(s(z, t))$ and corresponding

$$u(t) = \widehat{\gamma}^{-1}\left[-\widehat{\omega} + v(t) + u_{in}\right], \tag{4.41}$$

with $v(t)$ defined by (4.16) and

$$K \geq \frac{\delta + \widehat{\eta} + (1 - \beta)\,|v(t)|}{2 - \beta}, \tag{4.42}$$

$\widehat{\eta} > 0$ and $s(z, t) \in \mathbb{R}$ be a proposed novel sliding surface as

$$s(z, t) = (I_{t_s}^{1-\alpha} z_2)(t) + \Omega, \tag{4.43}$$

with $\dot{\Omega} = -v(t)$, leads to the establishment of finite time stabilization of uncertain chain of fractional order system (4.36), where n be the total number of fractional integrators and $t_f (t_s < t_f < \infty)$ the desired convergence time.

Proof Taking the Lyapunov candidate function as $V = \frac{1}{2}s^2$ (in place of $s(z, t)$, s is used for the simplicity of notation). The time derivative of candidate Lyapunov function is given as:

$$\dot{V} = s\dot{s}. \tag{4.44}$$

Using Eqs. (4.39), (4.43) and (4.40)

$$\begin{aligned} \dot{V} =& s\left((D_{t_s}^{\alpha} z_2)(t) + \dot{\Omega}\right) \\ =& s\left[\left(\Delta\omega - \Delta\gamma\widehat{\gamma}^{-1}\widehat{\omega}\right) + \left(1 + \Delta\gamma\widehat{\gamma}^{-1}\right)\widehat{u}(t) + \dot{\Omega}\right] \\ =& s\left[\left(\Delta\omega - \Delta\gamma\widehat{\gamma}^{-1}\widehat{\omega}\right) + \left(1 + \Delta\gamma\widehat{\gamma}^{-1}\right)v(t) - v(t)\right] \\ &- s\left[\left(1 + \Delta\gamma\widehat{\gamma}^{-1}\right)(K\operatorname{sign}(s))\right]. \end{aligned} \tag{4.45}$$

Using the inequality from (4.37) we can write

$$\dot{V} \leq s\left[\delta + (1 - \beta)v(t) - (2 - \beta)K\operatorname{sign}(s)\right]. \tag{4.46}$$

we select the gain K as in (4.42), so that the η-reachability condition of sliding mode is satisfied. One gets

$$s\dot{s} \leq -\widehat{\eta}\,\|s\|\,. \tag{4.47}$$

It means that if the inequality (4.42) is satisfied then (4.40) conforms the finite time reachability to s. During sliding $s = \dot{s} = 0$.

$$\begin{aligned} \dot{s} &= \left(\Delta\omega - \Delta\gamma\widehat{\gamma}^{-1}\widehat{\omega}\right) + \left(1 + \Delta\gamma\widehat{\gamma}^{-1}\right)\widehat{u}(t) + \dot{\Omega} = 0 \\ &= \left(\Delta\omega - \Delta\gamma\widehat{\gamma}^{-1}\widehat{\omega}\right) + \left(1 + \Delta\gamma\widehat{\gamma}^{-1}\right)\widehat{u}(t) - v(t) = 0 \\ &\Rightarrow \left(\Delta\omega - \Delta\gamma\widehat{\gamma}^{-1}\widehat{\omega}\right) + \left(1 + \Delta\gamma\widehat{\gamma}^{-1}\right)\widehat{u}(t) = v(t). \end{aligned} \tag{4.48}$$

Substituting in Eq. (4.39), the disturbance free dynamics is the same as given by (4.27) which becomes similar to (4.3), with n fractional order integrators. Hence a controller can be designed for the robust finite time steering of $x_1, x_2, \ldots, x_n$ from any initial point to origin.

4.5.1 Numerical Example

Consider the same system (4.28), (but here control part also contain uncertainties)

$$\begin{aligned} D_{t_s}^{\frac{1}{4}} x_1 &= x_2, \\ D_{t_s}^{\frac{1}{4}} x_2 &= \omega + \gamma u(t). \end{aligned} \tag{4.49}$$

Let $\omega = 2 + [-0.5 \;\; 0.5]$ and $\gamma = 1 + [-0.5 \;\; 0.5]$ satisfy the above assumption (which is given in this Section). Our aim is to calculate a switching control for the above system so that the system start at $t_s = 1$ with initial point $x(2, -3)$ to the final point $(0, 0)$ at time $t_f = 5$. Let the initial value of control be zero. Desired sliding surface for above goal is

$$s(x, t) = (I_{t_s}^{1-\alpha} x_2)(t) + \Omega, \quad \dot{\Omega} = -v(t). \tag{4.50}$$

where $v(t)$ is defined as in (4.35). Then, the control input $\widehat{u}(t)$ is the same as in (4.40). The δ and β must satisfy the following inequalities

$$\left|0.5 - \frac{0.5}{1}2\right| \leq \delta, \quad \left|\frac{0.5}{1}\right| \leq 1 - \beta, \tag{4.51}$$

and corresponding

$$u(t) = 1(-2 + v(t) - K\text{sign}(s)). \tag{4.52}$$

with

$$K \geq \frac{1.5 + \widehat{\eta} + (1 - 0.25)\,|v(t)|}{2 - 0.25}, \tag{4.53}$$

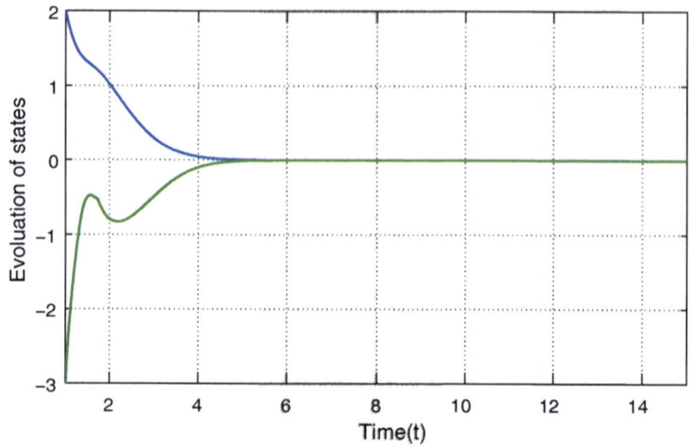

Fig. 4.2 Evolution of states, x_1 and x_2

here we assume $\beta = 0.25$, able to stabilizes the system (4.49) in finite time. As discussed in (4.48) applying u in (4.49), $s(x, t) = 0$ and the system (4.49) is converted into a similar form as given in (4.30). Now using $v(t)$ as derived in [21], sx_1 and x_2 are stabilized in finite time. Evolution of states are shown in Fig. 4.2 for $K = 5.5$.

4.6 Summary

In this chapter a novel methodology for finite time stabilization of a fractional order chain of integrator has been proposed for the first time. The finite time stabilization is achieved by using integral sliding like surface. The main contribution of this chapter is the design of finite time stabilizing robust controller using suitable integral sliding mode like surface such that the equivalent value of the integral sliding mode control is the negative of the disturbance and hence the disturbance is canceled. An improved strategy with more general kind of uncertainty is also proposed. The proposed controller can reject matched disturbances. Numerical examples are also presented to show the effectiveness of the proposed methods.

References

1. Utkin, V., Guldner, J., Shi, J.: Sliding mode control in electro-mechanical systems. Auton. Syst. **57**(1), 55–63 (2009)
2. Edwards, C., Spurgeon, S.K.: Sliding Mode Control: Theory and Applications, vol. 7, CRC Press, Boca Raton (1998)
3. Cao, W.J., Xu, J.X.: Nonlinear integral-type sliding surface for both matched and unmatched uncertain systems. IEEE Trans. Autom. Control **49**(8), 1355–1360 (2004)

4. Castaños, F., Fridman, L.: Analysis and design of integral sliding manifolds for systems with unmatched perturbations. IEEE Trans. Autom. Control **51**(5), 853–858 (2006)
5. Efe, M.Ö.: Fractional order sliding mode controller design for fractional order dynamic systems. In: Baleanu, D., Guvenc, Z.B., Machado, J.A.T. (eds.) New Trends in Nanotechnology and Fractional Calculus Application, pp. 463–470. Springer, Berlin (2010)
6. Efe, M.O.: Fractional order sliding mode control with reaching law approach. Turk. J. Electr. Eng. Comput. Sci **18**(5), 731–747 (2010)
7. Si-Ammour, A., Djennoune, S., Bettayeb, M.: A sliding mode control for linear fractional systems with input and state delays. Commun. Nonlinear Sci. Numer. Simul. **14**(5), 2310–2318 (2009)
8. Ramírez, H.S., Battle, V.F.: A generalized PI sliding mode and PWN control of switched fractional systems. In: Bartolini, G., Fridman, L., Pisano, A., Usai, E. (eds.) Modern Sliding Mode Control Theory, pp. 201–221. Springer, Berlin (2008)
9. Chang, Y.H., Wu, C.I., Chen, H.C., Chang, C.W., Lin, H.W.: Fractional-order integral sliding-mode flux observer for sensorless vector-controlled induction motors. In: American Control Conference (ACC) 2011, pp. 190–195, IEEE (2011)
10. HosseinNia, S.H., Sierociuk, D., Calderon, A.J., Vinagre, B.M.: Augmented system approach for fractional order smc of a dc-dc buck converter. In: Proceedings of IFAC Workshop on Fractional Differentiation and Its Applications, pp. 18–20 (2010)
11. Pisano, A., Rapaić, M., Usai, E.: Second-order sliding mode approaches to control and estimation for fractional order dynamics. In: Fridman, L., Moreno, J., Iriarte, R. (eds.) Sliding Modes after the First Decade of the 21st Century, pp. 169–197. Springer, Heidelberg (2012)
12. Ryan, E.: Singular optimal controls for second-order saturating systems. Int. J. Control **30**(4), 549–564 (1979)
13. Haimo, V.: Finite time controllers. SIAM J. Control Optim. **24**(4), 760–770 (1986)
14. Bhat, S.P., Bernstein, D.S.: Finite-time stability of continuous autonomous systems. SIAM J. Control Optim. **38**(3), 751–766 (2000)
15. Moulay, E., Perruquetti, W.: Finite time stability conditions for non-autonomous continuous systems. Int. J. Control **81**(5), 797–803 (2008)
16. Orlov, Y.: Finite time stability and robust control synthesis of uncertain switched systems. SIAM J. Control Optim. **43**(4), 1253–1271 (2004)
17. Moulay, E., Dambrine, M., Yeganefar, N., Perruquetti, W.: Finite-time stability and stabilization of time-delay systems. Syst. Control Lett. **57**(7), 561–566 (2008)
18. Lazarević, M.: Finite time stability analysis of pd¡ sup¿ α¡/sup¿ fractional control of robotic time-delay systems. Mech. Res. Commun. **33**(2), 269–279 (2006)
19. Nersesov, S.G., Haddad, W.M.: Finite-time stabilization of nonlinear impulsive dynamical systems. Nonlinear Anal. Hybrid Syst. **2**(3), 832–845 (2008)
20. Chen, W., Jiao, L.: Finite-time stability theorem of stochastic nonlinear systems. Automatica **46**(12), 2105–2108 (2010)
21. Andrzej, D., Wiktor, M.: Point to point control of fractional differential linear control systems. Adv. Differ. Equ. 2011, 13 (2011)
22. Kamal, S., Raman, A., Bandyopadhyay, B.: Finite-time stabilization of fractional order uncertain chain of integrator: an integral sliding mode approach. IEEE Trans. Autom. Control **58**(6), 1597–1602 (2013)
23. Chen, Y.Q., Petras, I., Xue, D.: Fractional order control-a tutorial. In: American Control Conference 2009, ACC'09, pp. 1397–1411. IEEE (2009)
24. Miller, K.S., Ross, B.: An Introduction to The Fractional Calculus and Fractional Differential Equations. John Wiley & Sons, INC (1993)
25. Anatoly, A.K., Srivastava, H.M., Trujillo, J.J.: Theory and Applictions of Fractional Differential Equations. Elsevier, Amsterdam (2006)

Chapter 5
A Soft Variable Structure Control of Fractional Order Systems

5.1 Introduction

Soft variable structure control [1–3], is a class of variable structure control lacking sliding modes, different from sliding mode VSC and switching supervisory control. In soft VSC, controller parameters or structures are continuously varying or switching. Due to continuous switching between different controllers it has many advantages such as achieving high regulation rates, shortening the arrival time and hardly any system chattering. We can refer to [1] and references within that for the detailed history, definition and unsolved problems of integer order soft VSC.

In soft variable structure actually Lyapunov function depends on the behavior of states and switch both. In Lyapunov stability analysis if the Lyapunov function and its derivative are quadratic, this results in exponential stability of the state, but here decrement is also dependent on switching strategy. Therefore, high regulation rate and short settling time comes into picture. Due to the above properties the term "fast asymptotic stability" is used, which is common in soft variable structure [1]. To the best of author's knowledge, soft VSC of fractional order system is yet to be reported [4].

In this Chapter, soft variable structure control of fractional order system is proposed for achieving high regulation rates and short settling times for the first time. Using the appropriate fractional order Lyapunov approach, bilinear switching based controller and the controller based on continuous fractional switch are designed to achieve high control performance. Due to the alleviation of chattering phenomenon soft VSC that is proposed for fractional order LTI systems will be quite useful in future for the practical systems.

This Chapter has been organized in the following way: Sect. 5.2 reviews the concept of fractional order extension of Lyapunov direct method. Main results of this Chapter are discussed in Sect. 5.3, in which Sect. 5.3.1 contains bilinear soft variable structure control. Section 5.3.2 contains dynamic soft variable structure control for fractional order system using differential fractional order switching function.

B. Bandyopadhyay and S. Kamal, *Stabilization and Control of Fractional Order Systems: A Sliding Mode Approach*, Lecture Notes in Electrical Engineering 317,
DOI 10.1007/978-3-319-08621-7_5

The robustness analysis of soft variable structure control for uncertain system using differential fractional order switching function is discussed in Sect. 5.3.3. Section 5.4 contains numerical results followed by the concluding Section.

5.2 Fractional Order Extension of Lyapunov Method

State space representation [5] of a fractional order linear time-invariant system is given as

$$\begin{aligned} {}_0D_t^{\alpha}x(t) &= Ax(t) + Bu(t), \\ y(t) &= Cx(t). \end{aligned} \tag{5.1}$$

where $x(t) \in \mathbb{R}^n$, $u(t) \in \mathbb{R}^r$ and $y(t) \in \mathbb{R}^p$ are states, input and output vectors of the system and $A \in \mathbb{R}^{n\times n}$, $B \in \mathbb{R}^{n\times r}$, $C \in \mathbb{R}^{p\times n}$ and $0 < \alpha < 1$ is the fractional commensurate order and the pair (A, B) is controllable. In this Chapter RL definition is considered henceforth unless otherwise specified.
In the proof of the main result, one can need the following remark, theorem and lemma.

Remark 5.1 The initial value condition for the ${}_0^{RL}D_t^{\alpha}x(t) = f(t,x)$ with $\alpha \in (0,1)$, $t > 0$ is $[{}_0^{RL}D_t^{\alpha-1}x(t)]_{t=0} = x_0'$ which can be replaced by $[t^{1-\alpha}x(t)]_{t=0} = \frac{x_0'}{\Gamma(\alpha)}$. However, the calculation of $[{}_0^{RL}D_t^{\alpha-1}x(t)]_{t=0} = x_0'$ is not feasible always. Above mathematical foundation is explained using the following Lemma

Lemma 5.2 *([6]) Let* $0 < \alpha < 1$ *and let* $x(t) \in \mathbb{C}_{1-\alpha}([0, b])$

(a) If

$$\lim_{t\to 0+}\left[t^{1-\alpha}x(t)\right] = c, \quad c \in \mathbb{R}. \tag{5.2}$$

Then

$$I^{1-\alpha}x(0^+) = \lim_{t\to 0^+} I^{1-\alpha}x(t) = c\Gamma(\alpha). \tag{5.3}$$

(b) If

$$\lim_{t\to 0^+} I^{1-\alpha}x(t) = b, \quad b \in \mathbb{R}, \tag{5.4}$$

and if there exists the limit, $\lim_{t\to 0^+}\left[t^{1-\alpha}x(t)\right]$, *then*

$$\lim_{t\to 0+}\left[t^{1-\alpha}x(t)\right] = \frac{b}{\Gamma(\alpha)}. \tag{5.5}$$

Same kind of transformation and brief history about the above transformation is also given in Chap. 3 of [6].

In the above mentioned theorem there is no restriction on $x(t)$ except for $x(t) \in C_{1-\alpha}([0, b])$, therefore validity of above mentioned transformation does not require any restriction on the behavior of Lyapunov function. Actually, this transformation gives one of the possible ways to take into account of the initial condition in some different co-ordinate frame.

Theorem 5.3 *Let $x = 0$ be an equilibrium point for the fractional non-autonomous system*

$$ {}_0D_t^{\alpha} x(t) = f(t, x). \tag{5.6}$$

Assume that there exists a Lyapunov function $V(t, x(t)) : [0, \infty) \times \mathbb{D} \to \mathbb{R}$ be a continuously differentiable function and locally Lipschitz with respect to x such that

$$\begin{aligned} \alpha_1 \|x\|^a \le V(t, x) \le \alpha_2 \|x\|^{ab}, \\ {}_0D_t^{\alpha} V(t, x) \le -\alpha_3 \|x\|^{ab}. \end{aligned} \tag{5.7}$$

where $t \ge 0$, $x \in \mathbb{D}$, $\alpha \in (0, 1)$, $\alpha_1, \alpha_2, \alpha_3, a$ and b are positive constants. Also assume that if $t^{1-\alpha} V(t, x)$ is continuous on $[0, T)$ where $T \to \infty$, and the fractional integral $I^{1-\alpha} V(t, x)$ of $V(t, x)$ is continuously differentiable for $t \in (0, T)$. Then $x = 0$ is Mittag-Leffler stable. If the assumptions hold globally on $\mathbb{R}^n$, then $x = 0$ is globally Mittag-Leffler stable.

Proof Using Eq. (5.7) one can write

$$ {}_0D_t^{\alpha} V(t, x) \le -\frac{\alpha_3}{\alpha_2} V(t, x). \tag{5.8}$$

There exists a nonnegative function $\eta(t, x) \in C((0, T) \times \mathbb{R}^n \to \mathbb{R})$ where $T \to \infty$ satisfying

$$\begin{aligned} {}_0D_t^{\alpha} V(t, x) + \eta(t, x) &= -\frac{\alpha_3}{\alpha_2} V(t, x), \\ \Rightarrow {}_0D_t^{\alpha} V(t, x) + \frac{\alpha_3}{\alpha_2} V(t, x) &= -\eta(t, x). \end{aligned} \tag{5.9}$$

Now using the assumption $t^{1-\alpha} V(t, x)$ is continuous on $[0, T)$ where $T \to \infty$, and the fractional integral $I^{1-\alpha} V(t, x)$ of $V(t, x)$ is continuously differentiable for $t \in (0, T)$, then Eq. (5.9), can be written as [7]

$$\begin{cases} {}_0D_t^{\alpha} V(t, x) + \frac{\alpha_3}{\alpha_2} V(t, x) = -\eta(t, x), \\ t^{1-\alpha} V(t)|_{t=0} = V(0), \quad t > 0. \end{cases} \tag{5.10}$$

One can note that differential equation involving Riemann-Liouville fractional derivative cannot have simple initial value condition as that of integer order, but the initial condition is given as $I^{1-\alpha} V(t)|_{t=0}$. One can further transfer initial value condition $I^{1-\alpha} V(t)|_{t=0}$ to $t^{1-\alpha} V(t)|_{t=0}$ and vice-versa [6].

If $x(0) = 0$, namely $V(0) = 0$, the solution to (5.10) is $x = 0$ and in this situation there exists a zero function $\eta(t, x) = 0$. If $x(0) \neq 0$, $V(0) > 0$, the above initial value problem has the following integral representation of solution [6, 7]

$$V(t) = \Gamma(\alpha)V(0)t^{\alpha-1}E_{\alpha,\alpha}\left(-\frac{\alpha_3}{\alpha_2}t^{\alpha}\right) - \eta(t, x) * \left[t^{\alpha-1}E_{\alpha,\alpha}\left(-\frac{\alpha_3}{\alpha_2}t^{\alpha}\right)\right], \quad (5.11)$$

where $*$ represents for the convolution operator and $E_{\alpha,\alpha}\left(-\frac{\alpha_3}{\alpha_2}t^{\alpha}\right)$ is Mittag-Leffler function [6]. Since both $t^{\alpha-1}$ and $E_{\alpha,\alpha}\left(-\frac{\alpha_3}{\alpha_2}t^{\alpha}\right)$ are nonnegative functions, one can further write

$$V(t) \leq \Gamma(\alpha)V(0)t^{\alpha-1}E_{\alpha,\alpha}\left(-\frac{\alpha_3}{\alpha_2}t^{\alpha}\right). \quad (5.12)$$

Now using (5.7) and (5.12), one can write

$$\|x(t)\| \leq \left[\frac{\Gamma(\alpha)V(0)t^{\alpha-1}}{\alpha_1}E_{\alpha,\alpha}\left(-\frac{\alpha_3}{\alpha_2}t^{\alpha}\right)\right]^{\frac{1}{a}}, \quad (5.13)$$

which imply the Mittag-Leffler stability of system (5.6). □

Lemma 5.4 [6] *The fractional integration operator* ${}_0I_t^{\alpha}$ *with fractional-order* α, *is bounded in* $\mathbb{L}_p(a_0, b_0)$, $(1 \leq p \leq \infty, -\infty < a_0 < b_0 < \infty)$

$$\|{}_0I_t^{\alpha} f(t)\| \leq \kappa_0 \|f(t)\|, \quad where\ \kappa_0 = \frac{(b_0 - a_0)^{\alpha}}{\alpha|\Gamma(\alpha)|}. \quad (5.14)$$

5.3 The Main Results

5.3.1 Bilinear Soft Variable Structure Control

In this Section the main result is discussed which is, stabilization of linear time invariant fractional order system (5.1) using the pair of linear subcontroller $u = L_{\Psi}^T x(t)$, which is switched between them such that Lyapunov function of the system (5.1) decreases as shown in Fig. 5.1.

For fulfilling the above mentioned goal a quadratic surface based switch is used, which takes the values either 1 or -1. The control equation for this approach is selected in such a way that, for $\Psi = 1$, L_2 will be active; for $\Psi = -1$, L_1 will be active. Thus

$$u = -\frac{1}{2}\left[(L_2 + L_1)^T + \Psi(L_2 - L_1)\right]x(t). \quad (5.15)$$

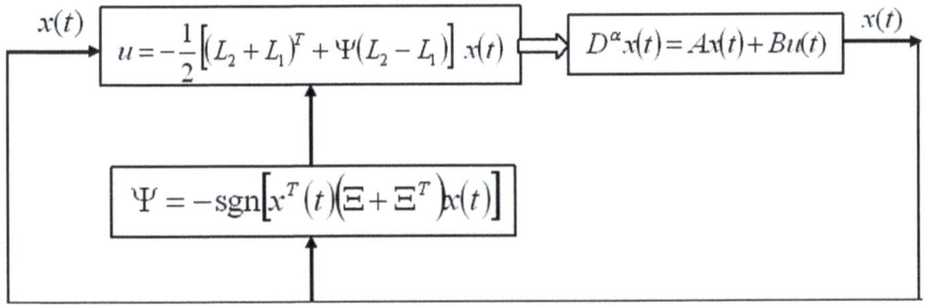

Fig. 5.1 Bilinear soft variable structure control

Inserting the control (5.15) into (5.1), the close loop system becomes

$$_0D_t^\alpha x(t) = \widehat{A}x(t) + \Psi \Xi x(t) = f_1(t, x). \tag{5.16}$$

which is bilinear in $\Psi \Xi$ and $x(t)$, where

$$\widehat{A} = A - \frac{1}{2}B(L_2 + L_1)^T, \quad \Xi = -\frac{1}{2}B(L_2 - L_1)^T. \tag{5.17}$$

The control vectors, L_2 and L_1, are chosen such that $\widehat{A}$ will be stable.

5.3.1.1 Design of Quadratic Surface Based Switch Ψ

Theorem 5.5 *The fractional order system (5.1) with control (5.15) and a selection strategy of the switch as*

$$\Psi = -sgn\left[x^T(t)(\Xi + \Xi^T)x(t)\right], \tag{5.18}$$

leads to fast asymptotic stability.

Proof Consider the candidate Lyapunov function as

$$V = 2x^T(t)x(t). \tag{5.19}$$

By using Leibniz's rule for fractional differentiation of V along the system (5.16), one gets

$$_0D_t^\alpha V = ({}_0D_t^\alpha x(t))^T x(t) + (x(t))^T({}_0D_t^\alpha x(t)) + 2W_1, \tag{5.20}$$

where

$$W_1 = \sum_{k=1}^{\infty} \frac{\Gamma(1+\alpha)}{\Gamma(1+k)\Gamma(1-k+\alpha)} \times \left({}_0D_t^k x(t)\right)^T \left({}_0D_t^{\alpha-k} x(t)\right). \tag{5.21}$$

As discussed in [8, 9]

$$\|W_1\| \le \Upsilon_1 \|x(t)\|^2, \tag{5.22}$$

where Υ_1 is a positive constant.

5.3.1.2 Calculation of Bound of Υ_1

One can write

$$\|W_1\| \le \sum_{k=1}^{\infty} \frac{\Gamma(1+\alpha)}{\Gamma(1+k)\Gamma(1-k+\alpha)} \times \|{}_0D_t^k x(t)\| \|{}_0D_t^{\alpha-k} x(t)\|. \tag{5.23}$$

After inserting the control (5.15) into (5.1), the right side of close loop system, $f_1(t,x) : [0,\infty) \times \mathbb{D} \to \mathbb{R}^n$ of (5.16) is piecewise continuous in t and $\mathbb{D} \in \mathbb{R}^n$ is a closed set that contains the origin $x = 0$. Since ${}_0D_t^k x(t)$ are piecewise continuous and also, bounded because $\mathbb{D}$ is the closed set. Thus, there exists $\bar{\rho}$ such that $\|{}_0D_t^k x(t)\| \le \bar{\rho}\|x\|$. Further, $\|{}_0D_t^{\alpha-k} x(t)\| \le \kappa_0\|x\|$ from the Lemma 5.4. Also, it is well known that the Gamma function [6] is nonzero everywhere along the real line. There is in fact no complex number z for which $\Gamma(z) = 0$, hence the reciprocal Gamma function $\frac{1}{\Gamma(z)}$ is an entire function, with zeros at $z = 0, -1, -2, \ldots$. Thus, there exists a bound $\hat{\rho}_{l0}$ such that $0 < \hat{\rho}_{l0} \le |\Gamma(1-\alpha+k)|$. Also, $\frac{\Gamma(k)}{\Gamma(k+1)} = \frac{1}{k}$, the infinite series $\sum_{k=1}^{\infty} \frac{1}{\Gamma(k+1)}$ converges. Therefore, there exists an upper bound $\hat{\rho}_{u0} > 0$ such that $0 < \sum_{k=1}^{\infty} \frac{1}{\Gamma(k+1)} < \hat{\rho}_{u0}$. Finally one can write Υ_1 as

$$\Upsilon_1 = \frac{\Gamma(1+\alpha)\bar{\rho}\kappa_0\hat{\rho}_{u0}}{\hat{\rho}_{l0}}. \tag{5.24}$$

Substituting (5.16) into (5.20)

$${}_0D_t^{\alpha} V = x^T(t)\left((\widehat{A} + \Psi\Xi) + (\widehat{A} + \Psi\Xi)^T\right)x(t) + 2W_1. \tag{5.25}$$

According to Theorem 5.3, for the stability of system (5.16)

$$x^T(t)\left(\widehat{A} + \widehat{A}^T\right)x(t) + \Psi x^T(t)\left(\Xi + \Xi^T\right)x(t) + 2W_1 < -\alpha_3\|x\|^2, \tag{5.26}$$

where $\alpha_3 > 0$. We now choose a positive definite matrix Q, such that

$$\widehat{A} + \widehat{A}^T = -Q, \tag{5.27}$$

and L_1, L_2 are selected such that

$$\lambda_{\min}\left\{\widehat{A} + \widehat{A}^T\right\} > 2\Upsilon_1, \tag{5.28}$$

so that the first term of Eq. (5.26) becomes negative. Using Rayleigh inequality

$$\lambda_{\min}\{Q\}\,\|x\|^2 \leq x^T Q x \leq \lambda_{\max}\{Q\}\,\|x\|^2. \tag{5.29}$$

One can write

$$\begin{aligned} {}_0D_t^{\alpha} V &\leq -\lambda_{\min}\{Q\}\,\|x\|^2 + 2\Upsilon_1\|x\|^2 + \Psi x^T(t)\left(\Xi + \Xi^T\right)x(t) \\ &\leq -\left(\lambda_{\min}\{Q\} - 2\Upsilon_1\right)\|x\|^2 + \Psi x^T(t)\left(\Xi + \Xi^T\right)x(t). \end{aligned} \tag{5.30}$$

Using $\Psi = -\mathrm{sgn}\left[x^T(t)(\Xi + \Xi^T)x(t)\right]$ as chosen in Eq. (5.18), the second term of (5.30) remains negative and given as

$$\begin{aligned} \Psi x^T(t)\left(\Xi + \Xi^T\right)x(t) &= -|x^T(t)\left(\Xi + \Xi^T\right)x(t)| \\ &\leq -\eta_1\|x\|^2, \end{aligned} \tag{5.31}$$

where η_1 is some positive constant. Using (5.30) and (5.31)

$${}_0D_t^{\alpha} V \leq -\widehat{\eta}_1\|x\|^2 - \eta_1\|x\|^2 \leq -(\widehat{\eta}_1 + \eta_1)\|x\|^2, \tag{5.32}$$

where $\widehat{\eta}_1 = (\lambda_{\min}\{Q\} - 2\Upsilon_1)$, which conforms the Mittag-Leffler stability of system (5.16). The main motive behind the choice of Ψ is to decrease Lyapunov function rapidly which leads to faster reaching of the trajectory to the equilibrium point. □

Remark 5.6 The problem that arises in the case of the bilinear variable structure control are that undesirable sliding modes might occur due to quadratic surface based switching function. In this case, motion is decoupled into slow and fast modes. Due to occurrence of slow motion our goal of achieving high regulation rate might not be satisfied.

In the next subsection a different strategy, called dynamic soft variable structure control scheme as shown in Fig. 5.2 is proposed, which is free from the above mentioned drawback.

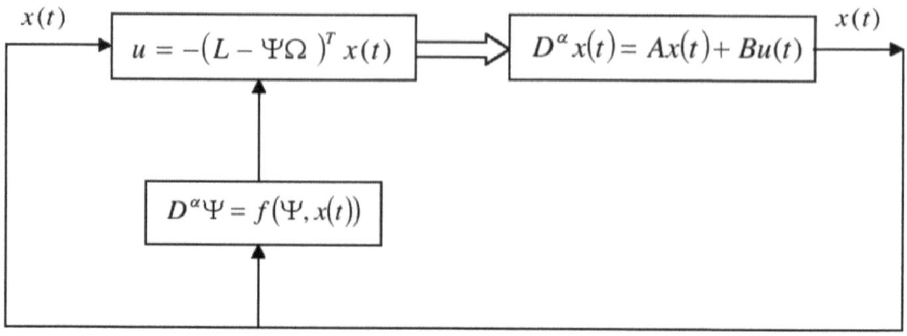

Fig. 5.2 Dynamic soft variable structure control

5.3.2 Dynamic Soft Variable Structure Control for the Fractional Order System Using Fractional Differential Switching Function

Let the selection switch of the following form

$$ {}_0D_t^{\alpha}\Psi = f(\Psi, x(t)). \tag{5.33}$$

Again considered the same linear fractional order plant (5.1) with allowing control constraints,

$$ |u| \le u_0. \tag{5.34}$$

Above control is bounded so only bounded set of initial state vector is considered here. Let the soft variable structure control for (5.1) is given by

$$ u = -(L - \Psi\Omega)^T x(t). \tag{5.35}$$

Closed loop fractional order system using above control (5.35) into (5.1)

$$ \begin{aligned} {}_0D_t^{\alpha}x(t) &= \left(A - BL^T + \Psi B\Omega^T\right)x(t) \\ &= \left(A_0 + \Psi B\Omega^T\right)x(t). \end{aligned} \tag{5.36}$$

where $A_0 = A - BL^T$. Using Eqs. (5.36) and (5.33), we obtain the following fractional differential system

$$ \begin{bmatrix} {}_0D_t^{\alpha}x(t) \\ {}_0D_t^{\alpha}\Psi \end{bmatrix} = \begin{bmatrix} (A_0 + \Psi B\Omega^T)x(t) \\ f(\Psi, x(t)) \end{bmatrix} \tag{5.37}$$

5.3.2.1 Switch Ψ for Dynamic Soft Variable Structure Control

Theorem 5.7 *For the fractional order system (5.37) with control (5.35) and selection strategy of the switch*

$$ {}_0D_t^{\alpha}\Psi = f(\Psi, x) = \frac{x^T(t)Q_2x(t) - \Psi\Theta(\Psi, x)}{q}, \tag{5.38} $$

with Q_2 as positive definite matrix and $\Theta(\Psi, x) > \widehat{\Upsilon}_2, q > 0$, where $\widehat{\Upsilon}_2 > 0$, leads to fast asymptotic stability.

Proof Consider the candidate Lyapunov function as

$$ V(\Psi, x) = 2x^T(t)x(t) + q\Psi^2, \tag{5.39} $$

where $q > 0$. By using Leibniz's rule for fractional differentiation of $V(\Psi, x)$ along the system (5.37), one gets

$$ \begin{aligned} {}_0D_t^{\alpha}V(\Psi, x) = &({}_0D_t^{\alpha}x(t))^T x(t) + (x(t))^T({}_0D_t^{\alpha}x(t)) \\ &+ 2\widehat{W}_1 + q({}_0D_t^{\alpha}\Psi)\Psi + W_2, \end{aligned} \tag{5.40} $$

where

$$ \widehat{W}_1 = \sum_{k=1}^{\infty} \frac{\Gamma(1+\alpha)}{\Gamma(1+k)\Gamma(1-k+\alpha)} \left({}_0D_t^{k}x(t)\right)^T \left({}_0D_t^{\alpha-k}x(t)\right), \tag{5.41} $$

and

$$ W_2 = q\sum_{k=1}^{\infty} \frac{\Gamma(1+\alpha)}{\Gamma(1+k)\Gamma(1-k+\alpha)} \left({}_0D_t^{k}\Psi\right) \left({}_0D_t^{\alpha-k}\Psi\right). \tag{5.42} $$

As discussed in [8, 9]

$$ \begin{aligned} \|\widehat{W}_1\| &\leq \widehat{\Upsilon}_1 \|x(t)\|^2, \\ \|W_2\| &\leq \widehat{\Upsilon}_2 \|\Psi\|^2, \end{aligned} \tag{5.43} $$

where $\widehat{\Upsilon}_1$ and $\widehat{\Upsilon}_2$ are positive constants. For the stability of system (5.37) fractional derivative of Lyapunov function must satisfied $({}_0D_t^{\alpha}V(\Psi, x) < -\alpha_3\|x\|^2 - \widehat{\alpha}_3\|\Psi\|^2)$

$$
\begin{aligned}
\Rightarrow {}_0D_t^{\alpha} V(\Psi, x) = x^T(t)\left(A_0 + \Psi B\Omega^T\right)^T x(t) \\
+ x^T(t)\left(A_0 + \Psi B\Omega^T\right) x(t) + 2\widehat{W}_1 \\
+ W_2 + q\Psi f(\Psi, x(t)).
\end{aligned} \tag{5.44}
$$

One can also write (5.44) as

$$
\begin{aligned}
{}_0D_t^{\alpha} V(\Psi, x) = x^T(t)\left(A_0 + A_0^T\right) x(t) \\
+ \Psi x^T(t)\left(B\Omega^T + \Omega B^T\right) x(t) + 2\widehat{W}_1 \\
+ W_2 + q\Psi f(\Psi, x(t)),
\end{aligned} \tag{5.45}
$$

$$
\begin{aligned}
\Rightarrow {}_0D_t^{\alpha} V(\Psi, x) = x^T(t)\left(A_0 + A_0^T\right) x(t) + 2\widehat{W}_1 \\
+ W_2 + \Psi\left(x^T(t)\left(B\Omega^T + \Omega B^T\right) x(t) + qf\right).
\end{aligned} \tag{5.46}
$$

For Mittag-Leffler stability of (5.37) (i.e., ${}_0D_t^{\alpha} V(\Psi, x) < -\alpha_3 \|x\|^2 - \widehat{\alpha}_3 \|\Psi\|^2$), we now choose

- Positive definite matrix Q_1 and Q_2,

$$A_0 + A_0^T = -Q_1. \tag{5.47}$$

$$B\Omega^T + \Omega B^T = -Q_2. \tag{5.48}$$

- Q_1, $\Theta(\Psi, x)$ such that

$$\lambda_{\min}\{Q_1\} > 2\widehat{\Upsilon}_1, \tag{5.49}$$

$$\Theta(\Psi, x) > \widehat{\Upsilon}_2, \tag{5.50}$$

- and

$$-x^T(t)Q_2x(t) + qf(\Psi, x) = -\Psi\Theta(\Psi, x), \tag{5.51}$$

then Eq. (5.46) yields

$$
\begin{aligned}
{}_0D_t^{\alpha} V(\Psi, x) = -x^T(t)Q_1x(t) + 2\widehat{W}_1 + W_2 \\
+ \Psi\left(-x^T(t)Q_2x(t) + qf\right).
\end{aligned} \tag{5.52}
$$

Using Eqs. (5.43), (5.49), (5.50), (5.51), (5.52) and Rayleigh inequality

$$\lambda_{\min}\{Q_1\}\|x\|^2 \le x^T Q_1 x \le \lambda_{\max}\{Q_1\}\|x\|^2. \tag{5.53}$$

One can write,

$$
\begin{aligned}
{}_0D_t^{\alpha} V(\Psi, x) \leq & -\lambda_{\min}\{Q_1\}\|x\|^2 - \Psi^2\Theta(\Psi, x) \\
& + 2\widehat{\Upsilon}_1\|x\|^2 + \widehat{\Upsilon}_2\|\Psi\|^2 \\
\leq & -\left(\lambda_{\min}\{Q_1\} - 2\widehat{\Upsilon}_1\right)\|x\|^2 \\
& - \|\Psi\|^2\left(\Theta(\Psi, x) - \widehat{\Upsilon}_2\right) \\
\leq & -\alpha_3\|x\|^2 - \hat{\alpha}_3\|\Psi\|^2,
\end{aligned}
\tag{5.54}
$$

where $\alpha_3 = \left(\lambda_{\min}\{Q_1\} - 2\widehat{\Upsilon}_1\right)$ and $\hat{\alpha}_3 = \left(\Theta(\Psi, x) - \widehat{\Upsilon}_2\right)$, which conforms the Mittag-Leffler stability of system (5.37). Also the presence of switching function leads to the decreasing of the Lyapunov function rapidly and trajectory reaches the equilibrium point faster.

For more clarity the above process can be explained as: Employing infinite number of subcontrollers, and by using a continuous fractional differential equation, incorporating a continuous function Ψ, the absolute values of the eigenvalues also continuously increases with Ψ, and the regulation rate will thus be greater, and the settling time be shorter [1–3]. Secondly, present form of control is not only responsible for the fast decrement of Lyapunov function but also contain one gain term $-BL^T$, which plays a significant role because, we design $A_0 = A - BL^T$ in such a way that it will exhibit better dynamic behavior, such as smoothly reaching to the origin and with a short arrival time.

Above process can also be explained mathematically in the following way:

Assume that the initial nonzero value of state $x(0) \neq 0$, is to be attenuated to $x(t) = 0$. At the beginning, we assume that the selection strategy $\Psi = 0$, then we can have $\Psi\Theta = 0$. In term of Eq. (5.38), we can obtain

$$
{}_0D_t^{\alpha}\Psi \approx \frac{x(t)^T Q_2 x(t)}{q}. \tag{5.55}
$$

As time t goes, we can deduce that $|\Psi|$ will increase and the derivative ${}_0D_t^{\alpha} V(\Psi, x)$ will decrease to point zero until ${}_0D_t^{\alpha} V(\Psi, x) \leq 0$, thus we get the Lyapunov function $V(\Psi, x)$ which will rapidly decrease and tend to the point zero, yielding a much rapid regulation process. After the accomplishment of reaching process the state vector x will decrease to a small neighborhood of zero. At this moment, the term $x^T(t)Q_2x(t)$ will thus approximately equal to zero in which case, we get

$$
{}_0D_t^{\alpha}\Psi \approx -\frac{\Psi\Theta(\Psi, x)}{q}, \tag{5.56}
$$

and $\Psi({}_0D_t^{\alpha}\Psi) < 0$. If parameter Ψ tends towards zero, the term $-BL^T$ will play a significant role because we design $A_0 = A - BL^T$ in such a way that it will exhibit better dynamics behavior, such as smoothly reaching to the origin and shortening the arrival time. Also L is designed such that it takes care of the actuator saturation problem. □

5.3.2.2 Constraints Imposed on Actuator Control Signals

Combining Eqs. (5.34) and (5.35), one gets

$$-u_0 \leq -L^T x + \Psi.\Omega^T x \leq u_0. \tag{5.57}$$

We convert Eq. (5.57) into

$$\begin{aligned} \frac{-u_0 + L^T x}{\Omega^T x} &\leq \Psi \leq \frac{u_0 + L^T x}{\Omega^T x} \quad \text{for } \Omega^T x > 0, \\ \frac{u_0 + L^T x}{\Omega^T x} &\leq \Psi \leq \frac{-u_0 + L^T x}{\Omega^T x} \quad \text{for } \Omega^T x < 0. \end{aligned} \tag{5.58}$$

Remark 5.8 As x approaches the equilibrium point $x = 0$, the selection variable Ψ take on very large values that can cause implementation problems and also amplify noise and disturbances in the actuator control signal. Hence an additional restrictions on Ψ is required. One can impose restriction as

$$-\Psi_1 \leq \Psi \leq \Psi_1. \tag{5.59}$$

Using Eqs. (5.58) and (5.59), one can get

$$\sigma(x) \leq \Psi \leq \omega(x), \tag{5.60}$$

where the σ and ω are the bounds depending on state vector x and given as

$$\sigma(x) = \begin{cases} \frac{u_0+L^T x}{\Omega^T x} & \text{for } \Omega^T x \leq \frac{u_0+L^T x}{\Psi_1} \\ -\Psi_1 & \text{for } \frac{u_0+L^T x}{\Psi_1} < \Omega^T x < \frac{-u_0+L^T x}{\Psi_1} \\ \frac{-u_0+L^T x}{\Omega^T x} & \text{for } \Omega^T x \geq \frac{-u_0+L^T x}{\Psi_1} \end{cases} \tag{5.61}$$

$$\omega(x) = \begin{cases} \frac{-u_0+L^T x}{\Omega^T x} & \text{for } \Omega^T x \leq \frac{-u_0+L^T x}{\Psi_1} \\ \Psi_1 & \text{for } \frac{-u_0+L^T x}{\Psi_1} < \Omega^T x < \frac{u_0+L^T x}{\Psi_1} \\ \frac{u_0+L^T x}{\Omega^T x} & \text{for } \Omega^T x \geq \frac{u_0+L^T x}{\Psi_1} \end{cases} \tag{5.62}$$

5.3.2.3 Suitable Choice for $\Theta(\Psi, x)$

$$\Theta(\Psi, x) = \begin{cases} \hbar\left(1 - \frac{\sigma(x)}{\Psi}\right) + \hbar_0\frac{\sigma(x)}{\Psi} & \text{for } \Psi \leq \sigma(x) \\ \hbar_0 & \text{for } \sigma(x) < \Psi < \omega(x) \\ \hbar\left(1 - \frac{\omega(x)}{\Psi}\right) + \hbar_0\frac{\omega(x)}{\Psi} & \text{for } \Psi \geq \omega \end{cases} \tag{5.63}$$

Remark 5.9 Equation (5.38) represents an anti-windup system. The constant $\hbar > 1$ determines the slope within the anti-windup region that prevents the value of integrators output parameter from falling below $\sigma(x)$ or exceeding $\omega(x)$. The constant $0 < \hbar_0 \leq 1$ determines the slope of region where the anti-windup is inactive.

5.3.3 Robustness Analysis of Dynamic Soft Variable Structure Control

Fractional differential system (5.37) with disturbance

$$\begin{bmatrix} {}_0D_t^\alpha x(t) \\ {}_0D_t^\alpha \Psi \end{bmatrix} = \begin{bmatrix} (A_0 + \Psi B\Omega^T)x(t) + g(t, x) \\ f(\Psi, x(t)), \end{bmatrix} \tag{5.64}$$

where $\|g(t, x)\| \leq \tilde{\gamma}_1\|x\|$ and $\|g(t, x)\| \leq \tilde{\gamma}_1\|x\| + \tilde{\gamma}_2$ in the case of vanishing and non-vanishing disturbance respectively for all $t \geq 0$ and all $x \in \mathbb{R}^n$. Same controller (5.35) and same switching strategy as stated in the Theorem 5.7 is applied to the perturbed system (5.64). Then one can easily prove the proposed controller is robust in the presence of disturbance also. Using same Lyapunov function (5.39) and using Leibniz's rule for fractional differentiation of $V(\Psi, x)$ along the system (5.64), one gets

$$\begin{aligned} {}_0D_t^\alpha V(\Psi, x) &= x^T(t)\left(A_0 + A_0^T\right)x(t) \\ &\quad + \Psi x^T(t)\left(B\Omega^T + \Omega B^T\right)x(t) + 2\widehat{W}_1 \\ &\quad + W_2 + q\Psi f(\Psi, x(t)) + g^T x(t) + x^T(t)g. \end{aligned} \tag{5.65}$$

For the stability of system (5.64) fractional derivative of Lyapunov function must satisfy (${}_0D_t^\alpha V(\Psi, x) < -\alpha_3\|x\|^2 - \widehat{\alpha}_3\|\Psi\|^2$). One can also write, (5.65) using (5.43), (5.47) and (5.51) as

$$\begin{aligned} {}_0D_t^\alpha V(\Psi, x) &\leq -\lambda_{\min}\{Q_1\}\|x\|^2 - \Psi^2\Theta(\Psi, x) \\ &\quad + 2\widehat{\Upsilon}_1\|x\|^2 + \widehat{\Upsilon}_2\|\Psi\|^2 + 2\|g\|\|x\|. \end{aligned} \tag{5.66}$$

In the case of vanishing disturbance one can write

$$
\begin{aligned}
{}_0D_t^{\alpha} V(\Psi, x) \leq & -\lambda_{\min}\{Q_1\}\|x\|^2 - \Psi^2\Theta(\Psi, x) \\
& + 2\widehat{\Upsilon}_1\|x\|^2 + \widehat{\Upsilon}_2\|\Psi\|^2 + 2\tilde{\gamma}_1\|x\|^2.
\end{aligned}
\tag{5.67}
$$

Therefore, if

$$
\lambda_{\min}\{Q_1\} > 2\widehat{\Upsilon}_1 + 2\tilde{\gamma}_1, \tag{5.68}
$$

and using (5.50)

$$
\begin{aligned}
{}_0D_t^{\alpha} V(\Psi, x) \leq & -\left(\lambda_{\min}\{Q_1\} - 2\widehat{\Upsilon}_1 - 2\tilde{\gamma}_1\right)\|x\|^2 \\
& - \|\Psi\|^2\left(\Theta(\Psi, x) - \widehat{\Upsilon}_2\right) = -\alpha_4\|x\|^2 - \widehat{\alpha}_3\|\Psi\|^2,
\end{aligned}
\tag{5.69}
$$

where $\alpha_4 = \lambda_{\min}\{Q_1\} - 2\widehat{\Upsilon}_1 + 2\tilde{\gamma}_1$ which conforms the Mittag-Leffler stability of the system (5.64). In the case of non-vanishing disturbance one can write

$$
\begin{aligned}
{}_0D_t^{\alpha} V(\Psi, x) \leq & -\lambda_{\min}\{Q_1\}\|x\|^2 - \Psi^2\Theta(\Psi, x) \\
& + 2\widehat{\Upsilon}_1\|x\|^2 + \widehat{\Upsilon}_2\|\Psi\|^2 + 2\tilde{\gamma}_1\|x\|^2 + 2\tilde{\gamma}_2\|x\|
\end{aligned}
\tag{5.70}
$$

Therefore, if

$$
\|x\| \geq \frac{2\tilde{\gamma}_2}{\lambda_{\min}\{Q_1\} - 2\widehat{\Upsilon}_1 - 2\tilde{\gamma}_1}. \tag{5.71}
$$

Then

$$
\begin{aligned}
{}_0D_t^{\alpha} V(\Psi, x) \leq & -\left(\lambda_{\min}\{Q_1\}\|x\| - 2\widehat{\Upsilon}_1\|x\|\right)\|x\| \\
& - (2\tilde{\gamma}_1\|x\| + 2\tilde{\gamma}_2)\|x\| \\
& - \|\Psi\|^2\left(\Theta(\Psi, x) - \widehat{\Upsilon}_2\right),
\end{aligned}
\tag{5.72}
$$

which implies that, in the presence of non-vanishing disturbance system trajectories remain ultimately bounded.

5.4 Numerical Results

Consider the following parameters for the single input ($u \in \mathbb{R}$) fractional order system (5.1)

$$
A = \begin{bmatrix} -1.2 & 0.6 & 1.2 \\ 0 & -0.8 & 3.6 \\ 0.1 & -0.2 & -0.5 \end{bmatrix}, \quad B = \begin{bmatrix} 0 \\ 0.8 \\ 0.5 \end{bmatrix}, \quad \alpha = 0.5. \tag{5.73}
$$

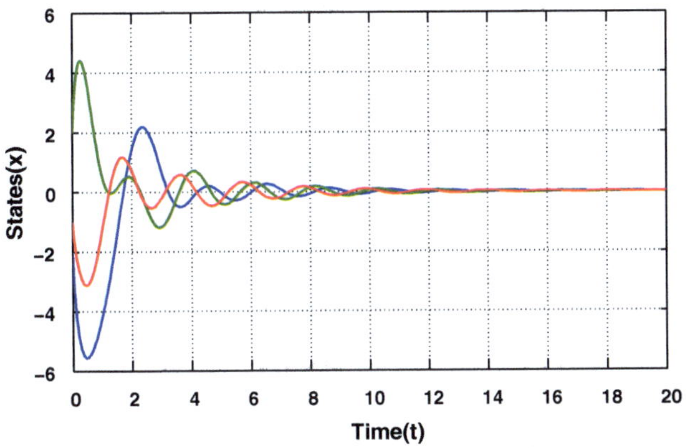

Fig. 5.3 Evolution of states without switching function ($\Psi = 0$)

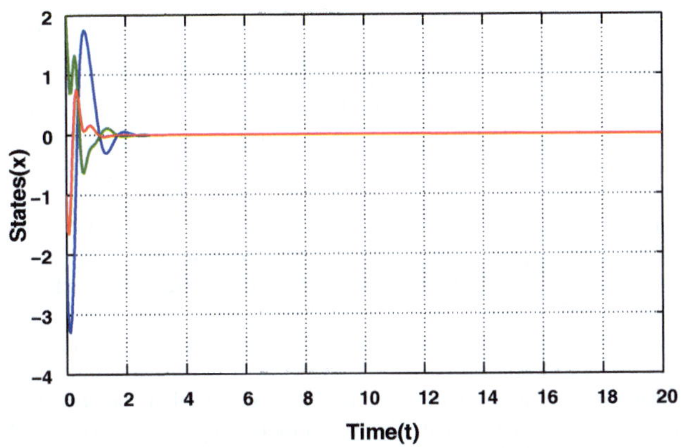

Fig. 5.4 Evolution of states with fractional order differential switch

The design parameters chosen are, $x(0) = [-2\ \ 2\ -1]$, $L^T = [1\ \ 2\ \ 3]$, $\Omega^T = [10\ \ 5\ \ 15]$, $\Theta = 100$, $q = 1{,}000$, $\hbar_0 = 10^{-2}$ and $\hbar = 10^6$, then $A_0 = A - BL^T$ which is given by

$$A_0 = \begin{bmatrix} -1.2 & 0.6 & 1.2 \\ -0.8 & -2.4 & 1.2 \\ -0.4 & -1.2 & -2.0 \end{bmatrix}. \tag{5.74}$$

Using (5.38) switching function is obtained and control law is obtained by Eq. (5.35). Evolution of states are shown in Figs. 5.3 and 5.4 with control containing no switching function $\Psi = 0$ and control containing fractional order differential switching function respectively. It is found from the simulation that the states of the fractional order system in the case of dynamic soft VSC are settled to equilibrium point quickly as compared to the simple state feedback.

5.5 Summary

In this Chapter, soft variable structure control of fractional order systems has been proposed for achieving high regulation rates and short settling times. This idea has been proposed for the first time. For obtaining high performance, a bilinear and continuous fractional switch based controller has been designed. Robustness analysis of dynamic soft variable structure control is also mathematically proved. Numerical examples are presented to illustrate the proposed method. Our future work will be the design of soft variable structure control for fractional order systems using an interger order switching function.

References

1. Adamy, J., Flemming, A.: Soft variable-structure controls: a survey. Automatica **40**(11), 1821–1844 (2004)
2. Su, J.P., Lee, T.E., Yu, K.W.: A combined hard and soft variable-structure control scheme for a class of nonlinear systems. IEEE Trans. Ind. Electron. **56**(9), 3305–3313 (2009)
3. Wang, Y., Xiong, Z., Ding, H.: Fast response and robust controller based on continuous eigenvalue configurations and time delay control. Robotics Comput. Integr. Manuf. **23**(1), 152–157 (2007)
4. Kamal, S., Bandyopadhyay, B.: High performance regulator for fractional order systems: A soft variable structure control approach. pp. xx-xx (2014 (Accepted))
5. Raynaud, H.F., Zergaınoh, A.: State-space representation for fractional order controllers. Automatica **36**(7), 1017–1021 (2000)
6. Anatoly, A.K., Srivastava, H.M., Trujillo, J.J.: Theory and applications of fractional differential equations, vol. 204 of (2006)
7. Zhang, S.: Monotone iterative method for initial value problem involving riemann-liouville fractional derivatives. Nonlinear Anal. Theory Methods Appl. **71**(5), 2087–2093 (2009)
8. Boroujeni, E.A., Momeni, H.R.: Observer based control of a class of nonlinear fractional order systems using lmi. World Acad. Sci. Eng. Technol. **61**, 779–782 (2012)
9. Dadras, S., Momeni, H.R.: A new fractional order observer design for fractional order nonlinear systems. In: Proceedings of the ASME 2011 International Design Engineering Technical Conference and Computers and Information in Engineering Conference, Washington, DETC2011-48861 (2011)

Chapter 6
Robust Cooperative Control of Fractional Order Multiple Agents

6.1 Introduction

Co-operative control of multiple agents have many applications in military services as well as in civil services. In cooperative control multiple entities work together to achieve certain common objective which depends on the type of application. Consensus is a common behavior excepted in most of the applications like rendezvous, formation control, flocking, target capturing etc. There is vast literature which talks about consensus in multiagent system where agents are modeled using integer order model [1–3]. But most of the systems in nature can be better modeled using fractional order differential equation rather than integer order, for example [4–6]:

- Chemotaxis behavior and food seeking of microbes and collective motion of bacteria in lubrications perspired by themselves.
- Underwater vehicles operating in lentic lakes composed of microbes and viscoelastic materials.
- Flying vehicles operating in an environment where the influence of particles in air cannot be ignored (e.g., high-speed flight in dust storm, rain, or snow).
- Ground vehicles moving on top of carpet, sand, muddy road or grass etc.,

Although distributed coordination of networked fractional-order systems over a directed interaction graph is recently reported in literature, there the agents are modeled using single fractional order integrator [4]. Also, sliding mode based algorithms to ensure distributed coordination of networked fractional-order systems over a directed interaction graph remains an unexplored field. To the best of the authors knowledge present work is the first attempt to address the more realistic problem. Main contributions of the chapter are

- Agents are modeled in a more realistic way using fractional order integrator.
- For achieving the specified goal a new fractional order control law is designed based on sliding mode theory.
- Finite time reachability to the sliding surface is proved using fractional order extension of Lyapunov stability criterion.
- Proposed controller is robust to matched uncertainties and continuous.

B. Bandyopadhyay and S. Kamal, *Stabilization and Control of Fractional Order Systems: A Sliding Mode Approach*, Lecture Notes in Electrical Engineering 317,
DOI 10.1007/978-3-319-08621-7_6

This chapter has been organized in the following way: In Sect. 6.2, some preliminaries of fractional order calculus and graph theory have been presented. The main result is presented in Sect. 6.4. Numerical example along with simulation is presented in Sect. 6.5. Finally, some concluding remarks are included.

6.2 Preliminaries

In this Section we introduce some basic concepts of the fractional order calculus and graph theory which is necessary for the development of further results of this chapter. ${}_0D_t^\alpha x(t)$ represents Caputo definition henceforth unless otherwise specified.

6.2.1 Fractional Order Calculus

For the proof of finite time reachability of the trajectory to the sliding surface and stability during sliding, one needs the following theorems and definition.

Theorem 6.1 [7] *The system ${}_0D_t^\alpha x(t) = Ax(t)$ is asymptotically stable if the following condition is satisfied*

$$|arg(eig(A))| > \frac{\alpha\pi}{2}, \tag{6.1}$$

where $0 < \alpha < 2$ and $eig(A)$ are eigenvalues of the matrix A.

Theorem 6.2 [8] *Let $x = 0$ be an equilibrium point for (either Caputo or Riemann-Liouville) non-autonomous fractional-order system*

$$ {}_0D_t^\alpha x(t) = f(t, x), \tag{6.2}$$

where $f(t, x)$ satisfies the Lipschitz condition with Lipschitz constant $l > 0$ and $\alpha \in (0, 1)$. Assume that there exist a Lyapunov candidate $V(t, x(t))$ satisfying

$$\begin{aligned} \alpha_1||x||^a \leq V(t, x) \leq \alpha_2||x||, \\ \dot{V}(t, x) \leq -\alpha_3||x||. \end{aligned} \tag{6.3}$$

where $\alpha_1, \alpha_2, \alpha_3$ and a are positive constants and $||.||$ denotes an arbitrary norm. Then the equilibrium point of the system (6.2) is Mittag-Leffer stable.

Remark 6.3 For $\alpha \in (0, 1)$ the Riemann Liouville and Caputo Derivative are related by

$$ {}_0^{RL}D_t^\alpha x(t) = {}_0D_t^\alpha x(t) + \frac{x(0)}{\Gamma(1-\alpha)t^\alpha}. \tag{6.4}$$

6.2.2 Graph Theory

Graphs provide the natural abstractions for how information is shared between agents in a network. A finite, undirected or simple graph is built upon a finite set, that is, a set that has a finite number of elements. Consider a system with n agents and let $\mathcal{G} = \{\mathcal{V}, \mathcal{E}\}$ represent the communication graph between the agents. $\mathcal{V}$ represents set of nodes where each node represent an agent and $\mathcal{E}$ represents set of edges. Following points are important to understand the graph theory

- The edge (i, j) in the edge set of a directed graph denotes that agent j can obtain information from agent i, but not necessarily vice versa.
- In the case of an undirected graph, edge (i, j) denotes that agents i and j can obtain information from each other. Therefore, one can interprets undirected graph as a special case of directed graph where an edge (i, j) in the undirected graph corresponds to the edges (i, j) and (j, i) in the directed graph.
- A weighted graph associates a weight with every edge in the graph.
- A directed graph is strongly connected if there is a directed path from every node to every other node. An undirected graph is connected if there is an undirected path between every pair of distinct nodes.
- An undirected graph is fully connected if there is an edge between every pair of distinct nodes.
- A directed graph is complete if there is an edge from every node to every other node.
- A directed tree is a directed graph in which every node has exactly one parent except for one node, called the root, which has no parent and which has directed paths to all other nodes.
- In undirected graphs, a tree is a graph in which every pair of nodes is connected by exactly one undirected path.
- A subgraph $\{\mathcal{V}^s, \mathcal{E}^s\}$ of $\{\mathcal{V}, \mathcal{E}\}$ is a graph such that $\mathcal{V}^s \subseteq \mathcal{V}$ and $\mathcal{E}^s \subseteq \mathcal{E} \cap (\mathcal{V}^s \times \mathcal{V}^s)$.

Theorem 6.4 [2] *The directed graph $\mathcal{G} = \{\mathcal{V}, \mathcal{E}\}$ has a directed spanning tree if and only if $\{\mathcal{V}, \mathcal{E}\}$ has at least one node with a directed path to all other nodes.*

Theorem 6.5 [2] *The Laplacian matrix L of a directed graph $\mathcal{G} = \{\mathcal{V}, \mathcal{E}\}$ has at least one zero eigenvalue and all of the nonzero eigenvalues are in the open right-half plane. In addition, L has exactly one zero eigenvalue if and only if $\mathcal{G}$ has a directed spanning tree. Furthermore, Rank$(L) = n$ if and only if L has a simple zero eigenvalue.*

6.3 Problem Statement

Consider a system of $n + 1$ agents described by fractional order dynamics, where an agent indexed by 0 acts as the target and other agents indexed by $1, \ldots, n$, are referred as the followers. The communication topology among the target and followers is described by a directed graph $\mathcal{G} = \{\mathcal{V}, \mathcal{E}\}$ with $\mathcal{V} = \{0, 1, \ldots, n\}$. The dynamics of the leader is described by the following fractional order integrator

$$\begin{aligned} {}_0D_t^\alpha x_{10} &= x_{20} \\ {}_0D_t^\alpha x_{20} &= u_0. \end{aligned} \tag{6.5}$$

The dynamics of the ith follower is described by the following fractional order integrator

$$\begin{aligned} {}_0D_t^\alpha x_{1i} &= x_{2i} \\ {}_0D_t^\alpha x_{2i} &= u_i + \rho_i. \end{aligned} \tag{6.6}$$

where ρ_i represents the matched disturbance and $u_i (i = 1, \ldots, n)$ is the control input to the ith agent.

In this development, it is considered that the x_{10} and x_{20} of the leader is available to its neighbors only and also the control input u_0 is unknown to any follower but its upper bound $|u_0|_{\max}$ is available to its neighbors. Let $\boldsymbol{\Phi}_1 := [e_1^1, \ldots, e_n^1]^T$ and $\boldsymbol{\Phi}_2 := [e_1^2, \ldots, e_n^2]^T$. The error functions e_i^1 and e_i^2 for agent i in case of consensus are defined as

$$e_i^1 := \sum_{j=1}^{n} a_{ij}(x_{1i} - x_{1j}) + b_i(x_{1i} - x_{10}), \tag{6.7}$$

$$e_i^2 := \sum_{j=1}^{n} a_{ij}(x_{2i} - x_{2j}) + b_i(x_{2i} - x_{20}). \tag{6.8}$$

If we want a specific formation of the follower agents with respect to the target, we can define error function e_i^1 as

$$e_i^1 := \sum_{j=1}^{n} a_{ij}(x_{1i} + \Delta_{1i} - x_{1j} - \Delta_{1j}) + b_i(x_{1i} + \Delta_{1i} - x_{10}). \tag{6.9}$$

Let A be the adjacent matrix defined as

$$A = \begin{pmatrix} 0 & 0 & \cdots & 0 \\ a_{10} & a_{11} & \cdots & a_{1n} \\ \vdots & \vdots & \ddots & \vdots \\ a_{n0} & a_{n1} & \cdots & a_{nn} \end{pmatrix}. \tag{6.10}$$

and

$$D = \operatorname{diag}\{d_0, d_2, \ldots, d_n\} \in \mathbb{R}^{(n+1)\times(n+1)}$$

be a diagonal matrix corresponding to graph $\mathcal{G}$ with $d_i = \sum_{j=1}^{n} a_{ij}$ for $i = 0, 1, 2, \ldots, n$. Then, the Laplacian of the graph $\mathcal{G}$ can be defined as

$$L = D - A.$$

Let $\bar{\mathcal{G}} = \{\bar{\mathcal{V}}, \bar{\mathcal{E}}\}$ be the subgraph of $\mathcal{G}$, which represents communication graph for the n followers, where

$$\bar{A} = \begin{pmatrix} a_{11} & a_{12} & \cdots & a_{1n} \\ \vdots & \vdots & \ddots & \vdots \\ a_{n1} & a_{n2} & \cdots & a_{nn} \end{pmatrix} \tag{6.11}$$

Let $\bar{A}$ be the adjacency matrix and

$$\bar{D} = \text{diag}\{\bar{d}_1, \bar{d}_2, ..., \bar{d}_n\} \in \mathbb{R}^{n \times n}$$

be a diagonal matrix corresponding to subgraph $\bar{\mathcal{G}}$ with $\bar{d}_i = \sum_{j=1}^{n} a_{ij}$ for $i = 1, 2, \ldots, n$. Then, the Laplacian of the graph $\bar{\mathcal{G}}$ can be defined as

$$\bar{L} = \bar{D} - \bar{A}.$$

The connection weight between agent i and the target is denoted by $\bar{B}$ where

$$\bar{B} = \text{diag}\{b_1, b_2, \ldots, b_n\} = \text{diag}\{a_{10}, a_{11}, \cdots, a_{1n}\}. \tag{6.12}$$

such that

$$b_i = \begin{cases} 1 & \text{if agent } i \text{ is connected to the target} \\ 0 & \text{otherwise} \end{cases} \tag{6.13}$$

Also, one can write

$$L = \begin{bmatrix} 0_{1\times n} & 0 \\ -[b_1 \ \cdots \ b_n]^T & [\bar{L} + \bar{B}] \end{bmatrix}.$$

The error dynamics of n agent system, where ith agents error dynamics are given by (6.7) and (6.8), can be expressed as

$$\begin{aligned} {}_0D_t^{\alpha}\boldsymbol{\Phi}_1 &= \boldsymbol{\Phi}_2 \\ {}_0D_t^{\alpha}\boldsymbol{\Phi}_2 &= (\bar{L} + \bar{B})U + (\bar{L} + \bar{B})\rho - \bar{B}\mathbf{1}u_0, \end{aligned} \tag{6.14}$$

where $U = [u_1, \ldots, u_n]^T$ and $\rho = [\rho_1, \ldots, \rho_n]^T$.

Our main aim is to design a sliding mode controller for the above MIMO system, so that $\boldsymbol{\Phi}_1 = \boldsymbol{\Phi}_2 = 0$ and simultaneously following objectives are satisfied

$$\begin{aligned} [x_{11}, \ldots, x_{1n}]^T &= \mathbf{1}x_{10} \\ [x_{21}, \ldots, x_{2n}]^T &= \mathbf{1}x_{20}, \end{aligned} \tag{6.15}$$

or

$$\begin{aligned} [x_{11}, \ldots, x_{1n}]^T &= \mathbf{1}x_{10} - [\Delta_1, \ldots, \Delta_n]^T \\ [x_{21}, \ldots, x_{2n}]^T &= \mathbf{1}x_{20}. \end{aligned} \tag{6.16}$$

where $\mathbf{1} \in \mathbb{R}^{n\times 1}$ represents the unity matrix. In the next section we present the design of continuous controller based on sliding mode.

6.4 Main Result

Sliding mode control design for a given system in general involves two steps. In the first step of design we select a specific manifold which depends on the type of application. Then we design a control such that all the trajectories of the system reach to the specified manifold in finite time and maintain the trajectories along the manifold. For achieving consensus in multiagent system described by (6.14), sliding surface is designed as

$$S = F\boldsymbol{\Phi}_1 + \boldsymbol{\Phi}_2 \tag{6.17}$$

where $F \in \mathbb{R}^{n\times n}$ is a diagonal matrix with every elements have positive value and $S = [s_1, \ldots, s_n]^T$. Taking the fractional order derivative of S along the (6.14), we can write

$$\begin{aligned} {}_0D_t^\alpha S &= F\, {}_0D_t^\alpha \boldsymbol{\Phi}_1 + {}_0D_t^\alpha \boldsymbol{\Phi}_2 \\ &= F\boldsymbol{\Phi}_2 + (\bar{L} + \bar{B})U + (\bar{L} + \bar{B})\rho - \bar{B}\mathbf{1}u_0. \end{aligned}$$

Substituting the value of $\boldsymbol{\Phi}_2$ from (6.17),

$${}_0D_t^\alpha S = FS - F^2\boldsymbol{\Phi}_1 + (\bar{L} + \bar{B})U + (\bar{L} + \bar{B})\rho - \bar{B}\mathbf{1}u_0. \tag{6.18}$$

Using (6.14), (6.17) and (6.18), we can write

$$\begin{aligned} {}_0D_t^\alpha \boldsymbol{\Phi}_1 &= S - F\boldsymbol{\Phi}_1 \\ {}_0D_t^\alpha S &= FS - F^2\boldsymbol{\Phi}_1 + (\bar{L} + \bar{B})U + (\bar{L} + \bar{B})\rho - \bar{B}\mathbf{1}u_0. \end{aligned} \tag{6.19}$$

6.4.1 Existence Condition of Sliding Mode

The main aim here, is to design U, such that the sliding motion occurs in finite time. For this purpose control is selected according to the following theorem.

Theorem 6.6 *Consider a multiagent system described by* (6.19). *Then the control input* U *defined as*

$$U = -(\bar{L} + \bar{B})^{-1}(-\nu + FS - F^2\boldsymbol{\Phi}_1), \tag{6.20}$$

where,

$$\nu = -(\eta + \varphi)\, {}_0I_t^{1-\alpha} \frac{S}{\|S\|_2}, \tag{6.21}$$

with $\eta > 0$, $\varphi > ||{}_0D_t^{1-\alpha}(\bar{L}+\bar{B})\rho - \bar{B}\mathbf{1}u_0|| + ||\frac{\nu(t=0)}{\Gamma(\alpha)t^{1-\alpha}}||$ *leads to the establishment of the sliding mode in finite time.*

Proof The closed loop system (6.19) with the control input (6.20) can be written as

$$\begin{aligned} {}_0D_t^{\alpha}\boldsymbol{\Phi}_1 &= S - F\boldsymbol{\Phi}_1 \\ {}_0D_t^{\alpha} S \ &= \nu + (\bar{L} + \bar{B})\rho - \bar{B}\mathbf{1}u_0. \end{aligned} \tag{6.22}$$

Now choosing the Lyapunov candidate function as $V = \frac{1}{2}S^T S = \frac{1}{2}||S||_2^2$ and taking the time derivative of V, along fractional order dynamical subsystem (6.22), we get

$$\begin{aligned} \dot{V} &= S^T \dot{S} \\ &= S^T {}_0D_t^{1-\alpha}({}_0D_t^{\alpha} S) \\ &= S^T {}_0D_t^{1-\alpha}(\nu + (\bar{L} + \bar{B})\rho - \bar{B}\mathbf{1}u_0). \end{aligned} \tag{6.23}$$

Using Eq. (6.4) and substituting the value of ν from (6.21),

$$\begin{aligned} \dot{V} &= S^T {}_0D_t^{1-\alpha}(\nu) + S^T {}_0D_t^{1-\alpha}((\bar{L} + \bar{B})\rho - \bar{B}\mathbf{1}u_0) \\ &= S^T {}_0^{RL}D_t^{1-\alpha}(\nu) - S^T \frac{\nu(t=0)}{\Gamma(\alpha)t^{1-\alpha}} + S^T {}_0D_t^{1-\alpha}((\bar{L} + \bar{B})\rho - \bar{B}\mathbf{1}u_0) \\ &= S^T {}_0^{RL}D_t^{1-\alpha}(-(\eta + \varphi){}_0I_t^{1-\alpha} \frac{S}{\|S\|_2}) \\ &\quad + S^T {}_0D_t^{1-\alpha}((\bar{L} + \bar{B})\rho - \bar{B}\mathbf{1}u_0) - S^T \frac{\nu(t=0)}{\Gamma(\alpha)t^{1-\alpha}}. \end{aligned} \tag{6.24}$$

$$\dot{V} = -(\eta + \varphi)S^T \frac{S}{\|S\|_2} + S^T \varphi_M - S^T \varphi_M', \tag{6.25}$$

where $\varphi_M = {}_0D_t^{1-\alpha}((\bar{L} + \bar{B})\rho - \bar{B}\mathbf{1}u_0)$ and $\varphi_M' = \frac{\nu(t=0)}{\Gamma(\alpha)t^{1-\alpha}}$. Applying the Cauchy-Schwartz inequality and simplifying (6.25) further, we can write

$$\dot{V} \leq -(\eta + \varphi)||S||_2 + \varphi||S||_2,$$
$$\leq -\eta||S||_2 \leq -\eta\sqrt{2V}. \tag{6.26}$$

Therefore, using Theorem 6.2, we can ensure the Mittag-Leffler stability of the sliding variable S which further ensures the finite time reaching of the trajectory to the sliding manifold. Now, we can calculate the time to reach the trajectory to the sliding manifold.

$$\dot{V} = \frac{dV}{dt} \leq -\eta\sqrt{2V}$$
$$t \leq \sqrt{2V_0}/\eta, \tag{6.27}$$

where V_0 is the initial value of Lyapunov function.

With the control input U as defined in (6.20), S can be reduced to zero in finite time. Then the reduced order dynamics of the system are given by,

$${}_0D_t^{\alpha}\boldsymbol{\Phi}_1 = -F\boldsymbol{\Phi}_1. \tag{6.28}$$

Now we can select F, according to the following Theorem.

Theorem 6.7 [7] *The system* ${}_0D_t^{\alpha}\boldsymbol{\Phi}_1 = -F\boldsymbol{\Phi}_1$ *is asymptotically stable if the following condition is satisfied*

$$|arg(eig(-F))| > \frac{\alpha\pi}{2}, \tag{6.29}$$

where $0 < \alpha < 2$ *and* $eig(-F)$ *are eigenvalues of the matrix* $-F$.

If condition (6.29) is satisfied the reduced order dynamics are stable. With the proper selection of F, we can make both $\boldsymbol{\Phi}_1$ and $\boldsymbol{\Phi}_2$ converge to the zero asymptotically. Following Theorem is necessary to established the convergence of the agents

Theorem 6.8 *Consider a multiagent system with agent dynamics* (6.6) *and target dynamics* (6.5). *If* $\boldsymbol{\Phi}_1 = 0$ *and the interconnection graph* $\mathscr{G}$ *has a directed spanning tree then the consensus is achieved.*

Proof When $\boldsymbol{\Phi}_1 = 0$,

$$(\bar{L} + \bar{B})[x_{11}, \ldots, x_{1n}]^T = \bar{B}\mathbf{1}x_{10} \tag{6.30}$$

As $\bar{L}\mathbf{1} = 0$, we can write

$$(\bar{L} + \bar{B})[x_{11}, \ldots, x_{1n}]^T = (\bar{L}\mathbf{1} + \bar{B}\mathbf{1})x_{10} = (\bar{L} + \bar{B})\mathbf{1}x_{10} \tag{6.31}$$

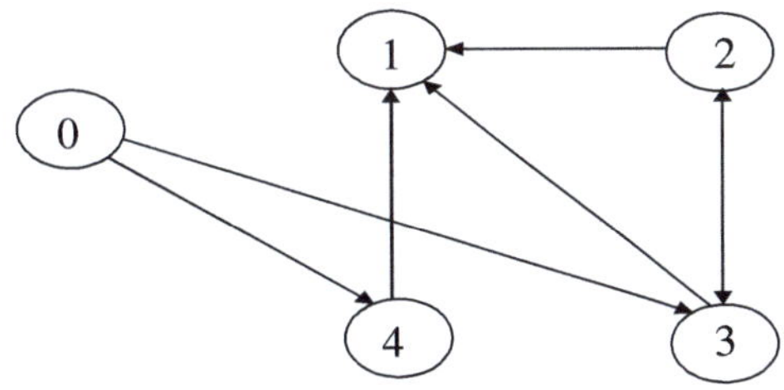

Fig. 6.1 Directed graph

The rank of laplacian L corresponding to graph $\mathscr{G}$ will be n as we are assuming that $\mathscr{G}$ has a directed spanning tree. We can write L as,

$$L = \begin{bmatrix} 0_{1\times n} & 0 \\ -[b_1 \ \cdots \ b_n]^T & [\bar{L} + \bar{B}] \end{bmatrix}. \tag{6.32}$$

As rank$(L) = n$ and $[\bar{L} + \bar{B}]\mathbf{1} = [b_1 \ \cdots \ b_n]^T$. So $[\bar{L} + \bar{B}]$ is invertible. Then using (6.31) we get,

$$[x_{11}, \ldots, x_{1n}]^T = \mathbf{1}x_{10} \tag{6.33}$$

which implies consensus to target trajectory. Similar results can be proved when we want to achieve target tracking with the desired formation about the target. In that case consensus is achieved if

$$[x_{11}, \ldots, x_{1n}]^T = \mathbf{1}x_{10} - [\Delta_1, \ldots, \Delta_n]^T. \tag{6.34}$$

That completes the proof of Theorem 6.8.

6.5 Simulation Results

We considered 4 homogeneous agents with random initial conditions. The agents are homogeneous in the sense that they all are represented with fractional order model having same α value. We assume fixed communication topology with the adjacency matrix A as shown in Fig. 6.1,

$$A = \begin{bmatrix} 0\,0\,0\,0\,0 \\ 0\,0\,1\,1\,1 \\ 0\,0\,0\,1\,0 \\ 1\,0\,1\,0\,0 \\ 1\,0\,0\,0\,0 \end{bmatrix} \tag{6.35}$$

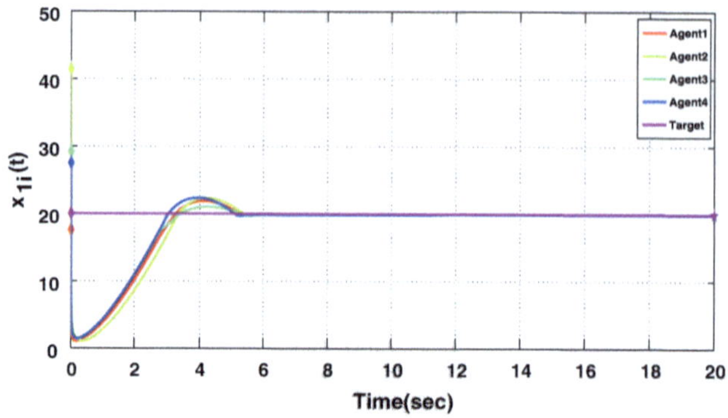

Fig. 6.2 Consensus for $\alpha = 0.5$

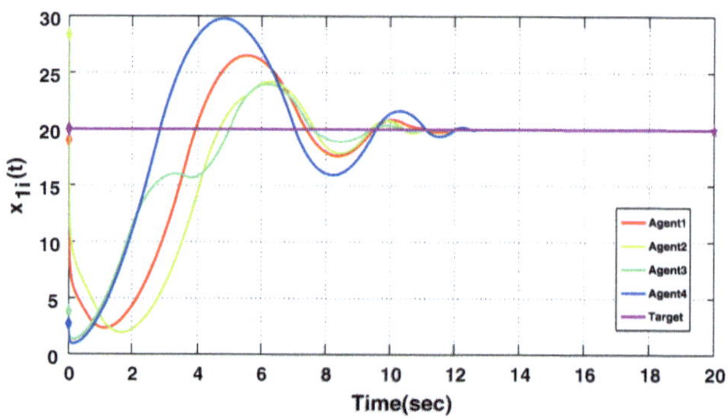

Fig. 6.3 Consensus for $\alpha = 0.8$

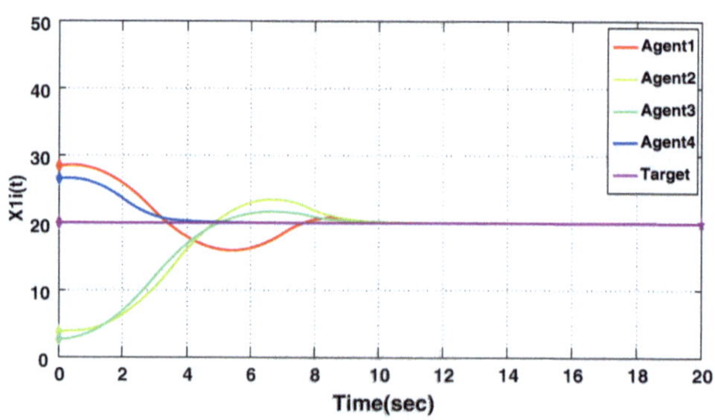

Fig. 6.4 Consensus for $\alpha = 1$

First we discuss a stationary target case. The target is initialized with $x_{10}(0) = 20$, $x_{20}(0) = 0$ and control input $u_0 = 0$. From the Figs. 6.2, 6.3 and 6.4, we observe that, the agents are able to follow the target. We considered different values for $\alpha = 0.5, 0.8$ and 1. When $\alpha = 1$ the results are same as those obtained with double

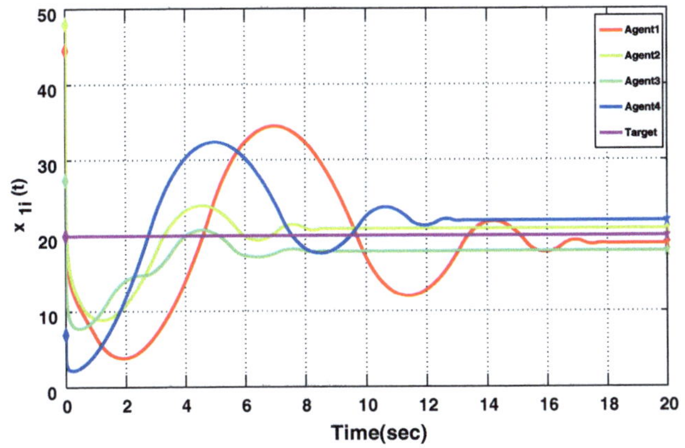

Fig. 6.5 Formation of agents

integrator model for each agent. To illustrate the formation control, we assumed desired relative distance with respect to the target as $\Delta_1 = 1$, $\Delta_2 = -1$, $\Delta_3 = 2$ and $\Delta_4 = -2$. As shown in the Fig. 6.5, the agents move along the target maintaining desired formation.

6.6 Summary

This chapter, investigates the problem of achieving distributed coordination of networked fractional order multi-agent systems over a directed interaction graph. For achieving the specified goal a new fractional order control law is designed based on sliding mode theory. Finite time reachability to the sliding surface is proved using fractional order extension of Lyapunov stability criterion. Effectiveness of the proposed scheme is illustrated with the simulation results.

References

1. Cao, Y., Yu, W., Ren, W., Chen, G.: An overview of recent progress in the study of distributed multi-agent coordination. IEEE Trans. Ind. Inf. **9**(1), 427–438 (2013)
2. Ren, W., Beard, R.: Distributed consensus in multi-vehicle cooperative control: theory and applications. Springer, Berlin (2007)
3. Ren, W., Cao, Y.: Distributed coordination of multi-agent networks: emergent problems, models, and issues. Springer, London (2010)
4. Cao, Y., Li, Y., Ren, W., Chen, Y.Q.: Distributed coordination of networked fractional-order systems. IEEE Trans. Syst. Man Cybern. Part B Cybern. **40**(2), 362–370 (2010)
5. Cohen, I., Golding, I., Ron, I., Ben-Jacob, E.: Biofluiddynamics of lubricating bacteria. Math. Methods Appl. Sci. **24**(17–18), 1429–1468 (2001)
6. Kozlovsky, Y., Cohen, I., Golding, I., Ben-Jacob, E.: Lubricating bacteria model for branching growth of bacterial colonies. Phys. Rev. E **59**(6), 7025–7035 (1999)

7. Petráš, I.: Stability of fractional-order systems with rational orders: a survey. Fractional Calc. Appl. Anal. **12**(3), 269–298 (2009)
8. Li, Y., Chen, Y., Podlubny, I.: Mittag-leffler stability of fractional order nonlinear dynamic systems. Automatica **45**(8), 1965–1969 (2009)

Chapter 7
Discrete Sliding Mode Control of Fractional Order Systems

7.1 Introduction

All the developments reported in the earlier chapters are for the continuous fractional order systems. However, a large class of continuous time systems are controlled by computers and micro-controllers. In these cases information about the system is available only at specific time instants and control inputs can only be applied at these instants. Due to this, discrete modeling of systems and controllers are more relevant than their continuous counterpart. Not much work is reported in the area of discrete-time fractional order modeling, stabilization and control. Although few results are available, they are applicable for restricted class of discrete fractional systems. These are discussed in the recent book [1], monograph [2] and references therein. State space representation of the linear discrete-time fractional order system has been introduced in [3]. Using this representation, observers, Kalman filters and feedback controllers are designed for the discrete fractional-order systems [4–7]. Some stability conditions have been derived in [8]. Observability and controllability properties of discrete fractional order system are discussed in [5].

Sliding mode controller [9] and Variable Structure Controllers (VSC) have been already implemented to improve the performance and robustness of closed loop control systems. Sliding Mode Control (SMC) has many attractive features such as invariance to matched uncertainties, model order reduction, simplicity in design, robustness against perturbations and some others [9–11]. The characteristic feature of continuous-time SMC system is that sliding mode occurs on a prescribed manifold (sliding surface), where switching control is employed to maintain the states on the surface [12].

To implement continuous-time variable structure control algorithm sampled-data environment is required, but in many practical places one cannot achieve performance one would expect on the basis of the continuous-time theory. Hence, in such a situation the option of discrete time sliding mode comes into picture. Also relatively low switching frequency is required than the continuous time sliding mode control (theoretically infinite frequency is required to maintain the trajectories on

B. Bandyopadhyay and S. Kamal, *Stabilization and Control of Fractional Order Systems: A Sliding Mode Approach*, Lecture Notes in Electrical Engineering 317,
DOI 10.1007/978-3-319-08621-7_7

the surface), so, discrete control algorithm is more practical to implement. The term 'discrete-time sliding mode' was first introduced by Utkin and Darkunov [13]. In case of Discrete Sliding Mode (DSM) design, the control input is computed only at certain sampling instants and the control effort is held constant over the entire sampling period.

The first work in the area of discrete variable structure control system was published in Russia in 1967. But more famous in discrete-time sliding mode control literature in the mid 80s by Milososavljevic [14], after that growing list of publication in this area such as, by Sarpturk et al. [15], Aly and All [16], Furuta [17], Sira-Ramirez [18], Spurgeon [19], Hung et al. [20], Bartolini et al. [21], Gao et al. [22], Jung and Tzou [23], Iordanou and Surgenor [24], Bartoszewicz [25], Hui and Zak [26], Golo and Milosavljevic [27], Koshkouei and Zinober [28].

The two schools of thought on discrete-time sliding mode control are (1) by using a switching based control law and (2) without using any switching function in the control law. In [22], Gao et al. has used the switching based reaching law, which shows that the motion of a discrete SMC system can undergo only quasi-sliding mode. It is also known that, chattering occurs in the system due to the application of a switching control [29]. It has been shown in [22], that due to the use of the switching function, the system states would reach the vicinity of the origin, but cannot get arbitrarily close to the origin. The other one is non switching type. It has been shown in [29] that, sliding mode may be achieved in discrete-time systems without the usage of a switching function. This is due to the fact that, discrete-time control is inherently discontinuous in nature and thus may not require an explicit discontinuity in the control law to bring out sliding mode control. Such a sliding mode control without chattering was proposed in [25]. Here, the control law uses apriori known function to lead the system states onto the sliding surface. The discrete-time sliding mode control using an equivalent control is proposed by Utkin [29]. The control does not need the switching function and it brings the state trajectory to the sliding surface in one sampling time. With limited control, trajectory reaches the sliding surface in finite sampling steps.

In the field of fractional order system control, the application of SMC was reported only recently in [1, 30, 31]. Asymptotic stabilization was achieved using higher order sliding mode control in [32]. However, it is limited to a continuous environment only. To the best of author's knowledge, sliding mode control of fractional discrete-time system is yet to be achieved.

In this chapter we propose a new methodology for the stabilization of fractional discrete-time systems based on discrete sliding mode approach for the first time [33]. Stability during sliding is analyzed in terms of gamma function based radius using the property introduced by Hilfer [34]. Since, sliding mode control is used, the proposed controller makes the system robust to matched uncertainties.

The chapter has been organized in the following way: In Sect. 7.2, some preliminaries of fractional discrete-time systems have been presented. The main result is presented in Sect. 7.3. In Sect. 7.4, designing of surface parameter and stability analysis is presented. In Sect. 7.5, quasi-sliding band for the fractional discrete time system containing matched uncertainty is obtained. Numerical example along with simulation is presented in the corresponding Sect. 7.6. Finally, some concluding remarks are included in Sect. 7.7.

7.2 Preliminaries of Fractional Discrete-Time Systems

For introducing the discrete time state space model of fractional order system, firstly discrete time equivalent of integer order is considered using the first forward difference, which is expressed as

$$\frac{x((k+1)h) - x(kh)}{h} = Ax(kh) + Bu(kh), \tag{7.1}$$

which yields the following approximate equivalent discrete state-space model for integer order system

$$\begin{aligned} x((k+1)h) &= (Ah + I)x(kh) + Bhu(kh) \\ y(kh) &= Cx(kh) + Du(kh), \end{aligned} \tag{7.2}$$

where $x \in \mathbb{R}^{n\times 1}$, $A \in \mathbb{R}^{n\times n}$, $B \in \mathbb{R}^{n\times m}$, $C \in \mathbb{R}^{p\times n}$, $D \in \mathbb{R}^{p\times m}$, $u(kh) \in \mathbb{R}^{m\times 1}$ and h is the sampling time. Now we will generalize the discretization method for fractional order system using the following Grunwald-Letnikov difference, which is defined as

Definition 7.1 The fractional order Grunwald-Letnikov difference of the function $x(k)$ is given by

$$\Delta^{\alpha} x(k) = \frac{1}{h^{\alpha}} \sum_{j=0}^{k} (-1)^j \binom{\alpha}{j} x(k-j), \tag{7.3}$$

where $\alpha \in \mathbb{R}$ is the order of fractional difference [35], $\mathbb{R}$ is the set of real numbers, h is the sampling interval and k is the number of samples for which the approximation of the derivatives is calculated. The binomial term $\binom{\alpha}{j}$ may be obtained from

$$\binom{\alpha}{j} = \begin{cases} 1 & \text{for } j = 0 \\ \frac{\alpha(\alpha-1)\ldots(\alpha-j+1)}{j!} & \text{for } j = 1, 2, 3, \ldots \end{cases} \tag{7.4}$$

Note Later assumed to be $h = 1$, then fractional $\alpha-$ order difference of the function $x(k)$ become

$$\Delta^{\alpha} x(k) = \sum_{j=0}^{k} (-1)^j \binom{\alpha}{j} x(k-j). \tag{7.5}$$

Definition 7.2 State space representation of the fractional discrete-time linear system is given as

$$\begin{aligned}\Delta^{\alpha}x(k+1) &= A_d x(k) + Bu(k)\\ x(k+1) &= \Delta^{\alpha}x(k+1) - \sum_{j=1}^{k+1}(-1)^j\binom{\alpha}{j}x(k-j+1)\\ y(k) &= Cx(k) + Du(k),\end{aligned} \tag{7.6}$$

where $x(k) \in \mathbb{R}^n$, $u(k) \in \mathbb{R}^m$, $y(k) \in \mathbb{R}^p$ are the state, input and output vectors and $A_d \in \mathbb{R}^{n\times n}$, $B \in \mathbb{R}^{n\times m}$, $C \in \mathbb{R}^{p\times n}$, $D \in \mathbb{R}^{p\times m}$.

Definition 7.3 State space representation of the nonlinear fractional discrete-time system is given as

$$\begin{aligned}\Delta^{\alpha}x(k+1) &= f(x(k), u(k))\\ x(k+1) &= \Delta^{\alpha}x(k+1) - \sum_{j=1}^{k+1}(-1)^j\binom{\alpha}{j}x(k-j+1)\\ y(k) &= h(x(k)),\end{aligned} \tag{7.7}$$

where the nonlinear functions $f(.)$ and $h(.)$, which are assumed to be of the class C^{∞}.

7.3 Main Result

The design of sliding mode control for the system (7.6) is discussed here.

Let the sliding surface for the fractional-order system (7.6) of the form:

$$s(k) = Sx(k). \tag{7.8}$$

Theorem 7.4 *System (7.6) is in discrete-time sliding mode if the $(SB)^{-1}$ is nonsingular and control $u(k)$ is designed as the solution of $s(k+1) = 0$, such that in one sampling instant trajectories come to the sliding surface and maintain there in the next successive instants.*

Proof Consider the linear reaching law

$$s(k+1) = \Phi s(k), \tag{7.9}$$

where $\Phi \in \mathbb{R}$ is a diagonal matrix with all entries between zero to one. This reaching law also satisfy the Sarpturk reaching law [15] $|s(k+1)| < |s(k)|$ and putting $\Phi = 0$

it is the same reaching law from which the equivalent control is obtained by Utkin [29]. Substituting $x(k+1)$ from (7.6) into $s(k+1) = 0$. One gets

$$\begin{aligned} s(k+1) &= Sx(k+1) = 0 \\ &= S\left(\Delta^{\alpha}x(k+1) - \sum_{j=1}^{k+1}(-1)^j \binom{\alpha}{j} x(k-j+1)\right) \\ &= S(A_d x(k) + Bu(k)) - S\left(\sum_{j=1}^{k+1}(-1)^j \binom{\alpha}{j} x(k-j+1)\right) = 0. \end{aligned} \tag{7.10}$$

and $u(k)$ is

$$u(k) = -\,(SB)^{-1}\,S\left(A_d x(k) + \sum_{j=1}^{k+1}(-1)^j \binom{\alpha}{j} x(k-j+1)\right). \tag{7.11}$$

Whenever matched uncertainties come into picture, it is easier to design sliding surface by converting system into regular form. Hence we discuss the regular form transformation of fractional order systems. Sliding surface design for systems, free from uncertainty is discussed in this section and quasi sliding mode band is obtained in Sect. 7.5, when matched uncertainty comes into picture.

Lemma 7.5 *There exists a nonsingular transformation matrix T, which defines the state transformation $z(k) = Tx(k)$ such that the system (7.6) is transformed into the regular form.*

$$\begin{aligned} \Delta^{\alpha}z(k+1) &= TA_dT^{-1}z(k) + TBu(k) \\ z(k+1) &= \Delta^{\alpha}z(k+1) - \sum_{j=1}^{k+1}(-1)^j \binom{\alpha}{j} z(k-j+1), \\ y(k) &= CT^{-1}x(k) + Du(k), \end{aligned} \tag{7.12}$$

where

$$TA_dT^{-1} = \begin{pmatrix} A_{11} & A_{12} \\ A_{21} & A_{22} \end{pmatrix} \quad \text{and} \quad TB = \begin{pmatrix} 0 \\ B_2 \end{pmatrix}, \tag{7.13}$$

also switching surface (7.8) becomes

$$s(k) = S_1 z_1(k) + S_2 z_2(k), \tag{7.14}$$

where $ST^{-1} = [S_1 \;\; S_2]$.

Using (7.12) and (7.13) one gets

$$\begin{aligned} \Delta^{\alpha} z_1(k+1) &= A_{11} z_1(k) + A_{12} z_2(k) \\ \Delta^{\alpha} z_2(k+1) &= A_{21} z_1(k) + A_{22} z_2(k) + B_2 u(k). \end{aligned} \tag{7.15}$$

$$\begin{aligned} z_1(k+1) &= \Delta^{\alpha} z_1(k+1) - \sum_{j=1}^{k+1} (-1)^j \binom{\alpha}{j} z_1(k-j+1) \\ z_2(k+1) &= \Delta^{\alpha} z_2(k+1) - \sum_{j=1}^{k+1} (-1)^j \binom{\alpha}{j} z_2(k-j+1). \end{aligned} \tag{7.16}$$

For calculating the state vector at $(k+1)$ instant, substituting fractional order difference vector (7.15) at $(k+1)$ instant into Eq. (7.16)

$$\begin{aligned} z_1(k+1) &= A_{11} z_1(k) + A_{12} z_2(k) - \sum_{j=1}^{k+1} (-1)^j \binom{\alpha}{j} z_1(k-j+1) \\ z_2(k+1) &= A_{21} z_1(k) + A_{22} z_2(k) - \sum_{j=1}^{k+1} (-1)^j \binom{\alpha}{j} z_2(k-j+1) + B_2 u(k). \end{aligned} \tag{7.17}$$

Theorem 7.6 *System (7.6) is in discrete-time sliding mode if the $(S_2 B_2)^{-1}$ is nonsingular and control $u(k)$ is designed as the solution of $s(k+1) = 0$, such that in one sampling instant trajectories come to the sliding surface and maintain there in the next successive instants.*

Proof Substituting $z_1(k+1)$ and $z_2(k+1)$ from (7.17) into $s(k+1) = 0$. One can gets

$$\begin{aligned} s(k+1) &= S_1 z_1(k+1) + S_2 z_2(k+1) = 0 \\ &= S_1 \left(A_{11} z_1(k) + A_{12} z_2(k)\right) - S_1 \left(\sum_{j=1}^{k+1} (-1)^j \binom{\alpha}{j} z_1(k-j+1) \right) \\ &\quad + S_2 \left(A_{21} z_1(k) + A_{22} z_2(k)\right) - S_2 \left(\sum_{j=1}^{k+1} (-1)^j \binom{\alpha}{j} z_2(k-j+1) + B_2 u(k) \right) \\ &= 0. \end{aligned} \tag{7.18}$$

and $u(k)$ is

$$\begin{aligned}u(k) =& -(S_2B_2)^{-1}S_1(A_{11}z_1(k)+A_{12}z_2(k))\\&+(S_2B_2)^{-1}S_1\left(\sum_{j=1}^{k+1}(-1)^j\binom{\alpha}{j}z_1(k-j+1)\right)\\&-(S_2B_2)^{-1}S_2(A_{21}z_1(k)+A_{22}z_2(k))\\&+(S_2B_2)^{-1}S_2\left(\sum_{j=1}^{k+1}(-1)^j\binom{\alpha}{j}z_2(k-j+1)\right).\end{aligned}\tag{7.19}$$

Remark 7.7 When linear reaching law (7.9) is used, then the above control (7.19) is modified as

$$\begin{aligned}u(k) =& (S_2B_2)^{-1}\Phi s(k)-(S_2B_2)^{-1}S_1(A_{11}z_1(k)+A_{12}z_2(k))\\&+(S_2B_2)^{-1}S_1\left(\sum_{j=1}^{k+1}(-1)^j\binom{\alpha}{j}z_1(k-j+1)\right)\\&-(S_2B_2)^{-1}S_2(A_{21}z_1(k)+A_{22}z_2(k))\\&+(S_2B_2)^{-1}S_2\left(\sum_{j=1}^{k+1}(-1)^j\binom{\alpha}{j}z_2(k-j+1)\right).\end{aligned}\tag{7.20}$$

7.4 Designing of Stable Sliding Surface

In this section the condition for the sliding surface parameter is obtained to make the reduced system stable during sliding mode.
After hitting the sliding surface $s(k)=0$ one can get

$$z_2(k)=-S_2^{-1}S_1z_1(k)=Mz_1(k),\tag{7.21}$$

where $M=-S_2^{-1}S_1$ and reduced order dynamics is given as

$$\begin{aligned}z_1(k+1) &= A_{11}z_1(k)+A_{12}Mz_1(k)-\sum_{j=1}^{k+1}(-1)^j\binom{\alpha}{j}z_1(k-j+1)\\&=(A_{11}+A_{12}M)z_1(k)-\sum_{j=1}^{k+1}(-1)^j\binom{\alpha}{j}z_1(k-j+1).\end{aligned}\tag{7.22}$$

The above system can be rewritten as an infinite dimensional system taking the form

$$\begin{bmatrix} z_1(k+1) \\ z_1(k) \\ z_1(k-1) \\ \dots \end{bmatrix} = A_{eq} \begin{bmatrix} z_1(k) \\ z_1(k-1) \\ z_1(k-2) \\ \dots \end{bmatrix}, \tag{7.23}$$

where

$$A_{eq} = \begin{bmatrix} (A_{11} + A_{12}M + \alpha I) & -(-1)^2 I \binom{\alpha}{2} & \cdots \\ I & 0 & \cdots \\ 0 & I & \cdots \\ \vdots & \vdots & \vdots \end{bmatrix}, \tag{7.24}$$

where I is identity operator.

Theorem 7.8 *Asymptotic stability of the above system (7.24) is given* $||A_{eq}|| < 1$, *where* $||.||$ *denotes the matrix norm defined as* $max|\lambda_i|$, *where* λ_i *is the* ith *eigenvalue of the matrix* A_{eq}.

Stability result for (7.22) like system is further simplified by Debeljkovic et al., [8] (considering the same as linear discrete-state delayed system).

The sliding surface is stable if the following condition satisfies (not an iff condition)

$$\|A_{11} + A_{12}M + \alpha I\| + \sum_{j=2}^{k+1} \left\| (-1)^j \binom{\alpha}{j} \right\| < 1. \tag{7.25}$$

For the real application the number of factors $(-1)^j \binom{\alpha}{j}$ in matrix A_{eq} has to be reduced. However this reduction may decrease the accuracy but system is still stable, only one has to ensure that the system is not close to the stability margin.

Considering the finite number of samples the above stability criteria is simplified as

$$\|A_{11} + A_{12}M + \alpha I\| + \sum_{j=2}^{L} \left\| (-1)^j \binom{\alpha}{j} \right\| < 1. \tag{7.26}$$

where L is finite number of samples. In order to make the above criterion more useful, a further simplification of (7.26) is presented by Debeljkovic et al. [8] using the following lemma.

Lemma 7.9 *[8] For* $j \geq 2$, *the factors* $(-1)^j \binom{\alpha}{j}$ *are*

$$(-1)^j \binom{\alpha}{j} = \begin{cases} positive & for\ 2 > \alpha > 1 \\ negative & for\ 0 < \alpha < 1 \\ 0 & for\ \alpha = 0, 1 \end{cases} \tag{7.27}$$

Theorem 7.10 *[8] The reduced order system given by (7.22) is stable if (but, not only if)*

$$\|A_{11} + A_{12}M + \alpha I\| < r(L, \alpha). \tag{7.28}$$

where

$$r(L,\alpha) = \begin{cases} 2 - \frac{\Gamma(L+1-\alpha)}{\Gamma(1-\alpha)\Gamma(L+1)} - \alpha & \text{for } \alpha \in \langle 1, 2 \rangle \\ \frac{\Gamma(L+1-\alpha)}{\Gamma(1-\alpha)\Gamma(L+1)} + \alpha & \text{for } \alpha \in \langle 0, 1 \rangle \\ 1 & \text{for } \alpha = 0 \end{cases} \tag{7.29}$$

and $r(L, \alpha)$ is a stability radius of the system, i.e., it is the radius of a circle within which stable eigenvalues of the system are to be located.

Note:- Actually

$$\sum_{j=0}^{L} (-1)^j \binom{\alpha}{j} = \frac{\Gamma(L+1-\alpha)}{\Gamma(1-\alpha)\Gamma(L+1)}$$

and

$$\sum_{j=2}^{L} (-1)^j \binom{\alpha}{j} = \frac{\Gamma(L+1-\alpha)}{\Gamma(1-\alpha)\Gamma(L+1)} - 1 + \alpha.$$

The above expression is derived by Hilfer [34].

Remark 7.11 Similarly for the equivalent control like (7.11) the stability radius is given by

$$\left\| A_d + \widetilde{M} A_d + \widetilde{M} \alpha I \right\| < \widetilde{M} r(L, \alpha). \tag{7.30}$$

where $\widetilde{M} = -B(SB)^{-1}S$.

7.5 Quasi-Sliding Band for the Fractional Discrete-Time System Containing Matched Uncertainty

State based feedback controller design for LTI-fractional order system in [4] is not robust with matched uncertainty. But one can show that in the presence of matched uncertainty the controller proposed in this chapter is robust and the state trajectory remains in band called quasi-sliding mode band. The band calculation is given as follows:

Equation (7.17) containing matched uncertainty can be written as

$$z_1(k+1) = A_{11}z_1(k) + A_{12}z_2(k) - \sum_{j=1}^{k+1}(-1)^j\binom{\alpha}{j}z_1(k-j+1)$$
$$z_2(k+1) = A_{21}z_1(k) + A_{22}z_2(k) - \sum_{j=1}^{k+1}(-1)^j\binom{\alpha}{j}z_2(k-j+1) + B_2u(k) + f_m. \tag{7.31}$$

where $f_m = B_2 f = [f_{m1}, f_{m2}, \cdots, f_{mm}]^{\mathrm{T}}$ is matched uncertainty.

Theorem 7.12 *System (7.31) under the control of (7.20) will be driven arbitrarily close to the quasi sliding mode band* ε_m

$$\varepsilon_m = \sqrt{\sum_{i=1}^{m}\left(\frac{1}{1-\Phi_i}S_2\widehat{f_{mi}}\right)^2}, \tag{7.32}$$

where $\widehat{f_{mi}} = max\{f_{mi}\}$.

Proof Using (7.14), (7.20) and (7.31), one can write

$$\begin{aligned} s(k+1) &= S_1z_1(k+1) + S_2z_2(k+1) \\ &= \Phi s(k) + S_2 f_m. \end{aligned} \tag{7.33}$$

where $\Phi = \mathrm{diag}\{\Phi_1, \Phi_2, \cdots, \Phi_m\}$. If control $u \in \mathbb{R}^m$, then m independent sliding surface $s = [s_1, s_2, \cdots, s_m]^{\mathrm{T}}$ is possible. If we start from $s_i(0)$, where $i = 1, 2, \cdots, m$ the future values of $s_i(k)$ can be found from the solution of (7.33)

$$s_i(k) = \Phi_i^k s_i(0) + \sum_{i=0}^{k-1}\Phi_i^k S_2 f_{mi}. \tag{7.34}$$

If $f_{mi} < \widehat{f_{mi}}$, then Eqn. (7.34) can be written as

$$|s_i(k)| < \left|\Phi_i^k s_i(0)\right| + \sum_{j=0}^{k-1}\Phi_i^k S_2\widehat{f_{mi}}. \tag{7.35}$$

For sufficiently large k the right hand side of the above inequality approaches towards $\varepsilon_i = \frac{1}{1-\Phi_i}S_2\widehat{f_{mi}}$. For each sliding surface similar relation is obtained after taking large value of k. One can calculate the average value of quasi-sliding band by taking Euclidean norm of all ε, which is same as (7.32). It is also obvious that for $\Phi = 0$ leads to $\varepsilon = \left\|S_2\widehat{f_{mi}}\right\|$, where $\widehat{f_{mi}} = \max\{f_{mi}\}$.

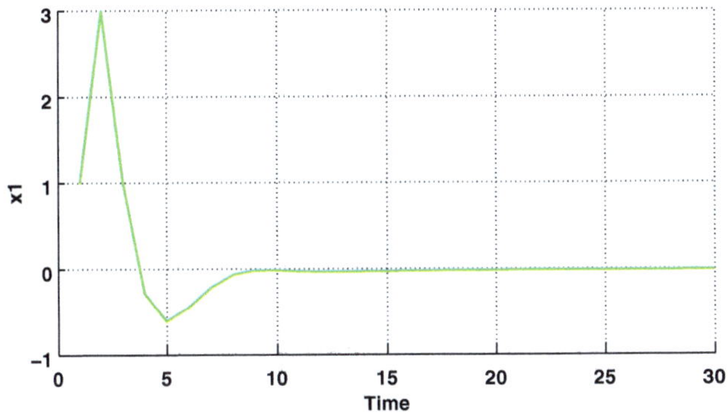

Fig. 7.1 Evolution of state(x1) w.r.t. time

7.6 Numerical Result

As an illustration of the results presented in this chapter, a discrete sliding mode controller for Ultracapacitor is designed. Consider the continuous fractional order state-space model of Ultracapacitor, which is given by

$$\frac{d^{\alpha}x}{dt^{n}} = Ax(t) + bu(t)$$
$$y == cx(t) + du(t). \tag{7.36}$$

Discrete fractional order state-space model given in [36] is written here as

$$A_d = \begin{bmatrix} 0 & 1 \\ 0.035311 & 0.001815 \end{bmatrix}, \quad B = \begin{bmatrix} 0 \\ 1 \end{bmatrix}, \quad \alpha = 0.2$$
$$C = \begin{bmatrix} -0.018624 & 0.188432 \end{bmatrix}, \quad D = \begin{bmatrix} 0 \end{bmatrix}, \quad T_s = 0.1 \tag{7.37}$$

For simulation, the initial conditions of state vector were chosen as $x1 = 1$, $x2 = 2$ and sliding surface parameter $s = [0.9 \quad 1.5]$. The numerical simulation for the first state ($x1$) is depicted in Fig. 7.1 and the second state ($x2$) is shown in Fig. 7.2. The Fig. 7.3 shows that the sliding surface which conforms that it becomes zero after one sampling time. Finally control effort for stabilization of states is shown in Fig. 7.4.

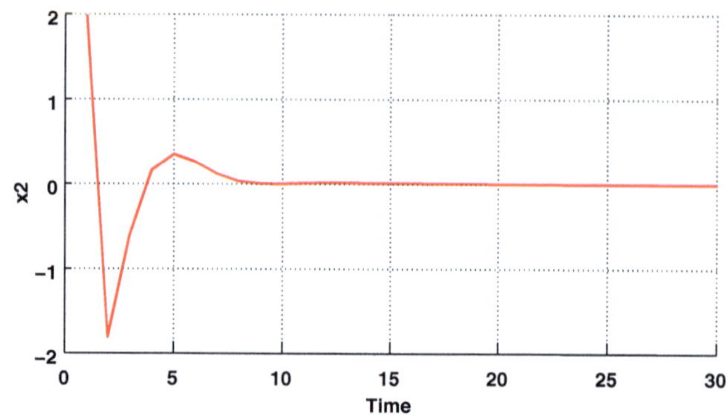

Fig. 7.2 Evolution of state(x2) w.r.t. time

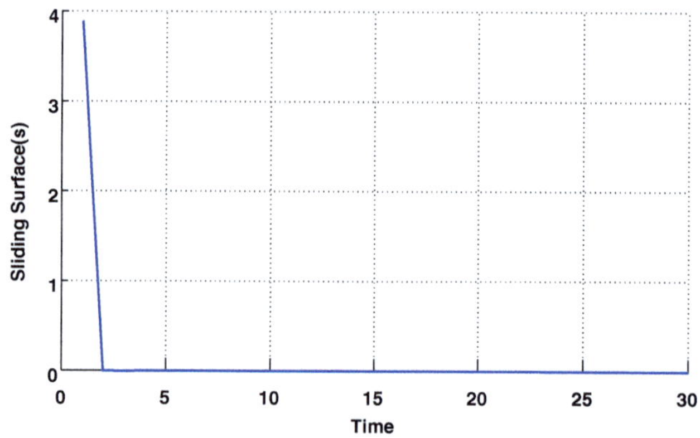

Fig. 7.3 Evolution of sliding surface(s) w.r.t. time

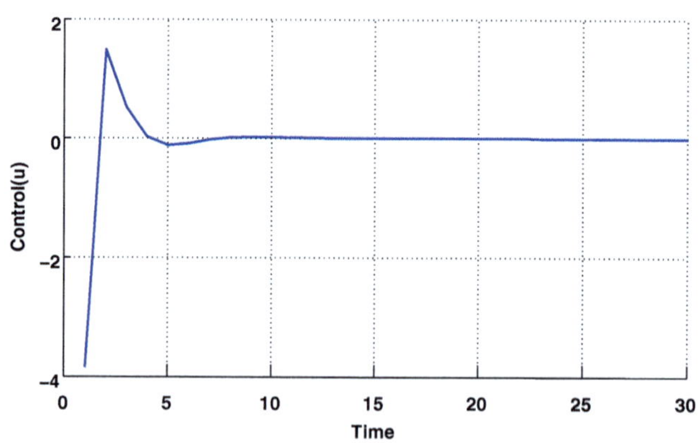

Fig. 7.4 Evolution of control(u(k)) w.r.t. time

7.7 Summary

In this chapter, a methodology for the stabilization of fractional discrete-time systems based on discrete sliding mode approach has been proposed for the first time. The proposed controller is robust in the presence of matched uncertainties. Stability during sliding has been analyzed in terms of gamma function based radius using the property introduced by Hilfer [34]. Numerical example along with simulation results have been presented to illustrate the efficacy of the proposed method.

References

1. Baleanu, D., Güvenç, Z.B., Machado, J.T. (eds.): New trends in nanotechnology and fractional calculus applications. Springer, Netherlands (2010)
2. Kaczorek, T.: Selected problems of fractional systems theory. Springer, Berlin (2011)
3. Raynaud, H.F., Zergaınoh, A.: State-space representation for fractional order controllers. Automatica **36**(7), 1017–1021 (2000)
4. Dzielinski, A., Sierociuk, D.: Adaptive feedback control of fractional order discrete state-space systems. In: International Conference on Computational Intelligence for Modelling, Control and Automation, 2005 and International Conference on Intelligent Agents, Web Technologies and Internet Commerce, vol. 1, pp. 804–809. (2005).
5. Guermah, S., Djennoune, S., Bettayeb, M.: Controllability and observability of linear discrete-time fractional-order systems. Int. J. Appl. Math. Comput. Sci. **18**(2), 213–222 (2008)
6. Podlubny, I.: Fractional-order systems and pi/sup/spl lambda//d/sup/spl mu//-controllers. IEEE Trans. Autom. Control **44**(1), 208–214 (1999)
7. Sierociuk, D., Dzieliński, A.: Fractional kalman filter algorithm for the states, parameters and order of fractional system estimation. Int. J. Appl. Math. Comput. Sci. **16**, 129–140 (2006)
8. Dzieliński, A., Sierociuk, D.: Stability of discrete fractional order state-space systems. J. Vib. Control **14**(9–10), 1543–1556 (2008)
9. Utkin, V., Guldner, J.: Sliding mode control in electro-mechanical systems. CRC Press, US (2009)
10. Edwards, C., Spurgeon, S.K.: Sliding mode control: theory and applications. CRC Press, US (1998)
11. Utkin, V.I.: Sliding modes in control and optimization. Springer, Berlin (1992)
12. Fridman, L., Levant, A.: Higher order sliding modes. Sliding Mode Control Eng. **11**, 53–102 (2002)
13. Drakunov, S.V., Utkin, V.: Sliding mode control in dynamic systems. Int. J. Control **55**(4), 1029–1037 (1992)
14. Milosavljevic, C.: General conditions for the existence of a quasi-sliding mode on the switching hyperplane in discrete variable structure systems. Autom. Remote Control **46**(3), 307–314 (1985)
15. Sarpturk, S., Istefanopulos, Y., et al.: On the stability of discrete-time sliding mode control systems. IEEE Trans. Autom. Control **32**(10), 930–932 (1987)
16. Aly, G.M., Ali, W.G.: Digital design of variable structure control systems. Int. J. Syst. Sci. **21**(8), 1709–1720 (1990)
17. Furuta, K.: Sliding mode control of a discrete system. Syst. Control Lett. **14**(2), 145–152 (1990)
18. Sira-Ramirez, H.: Non-linear discrete variable structure systems in quasi-sliding mode. Int. J. Control **54**(5), 1171–1187 (1991)
19. Spurgeon, S.: Hyperplane design techniques for discrete-time variable structure control systems. Int. J. Control **55**(2), 445–456 (1992)

20. Hung, J.Y., Gao, W., Hung, J.C.: Variable structure control: a survey. IEEE Trans. Industr. Electron. **40**(1), 2–22 (1993)
21. Bartolini, G., Ferrara, A., Utkin, V.I.: Adaptive sliding mode control in discrete-time systems. Automatica **31**(5), 769–773 (1995)
22. Gao, W., Wang, Y., Homaifa, A.: Discrete-time variable structure control systems. IEEE Trans. Industr. Electron. **42**(2), 117–122 (1995)
23. Jung, S.L., Tzou, Y.Y.: Discrete sliding-mode control of a pwm inverter for sinusoidal output waveform synthesis with optimal sliding curve. IEEE Trans. Power Electron. **11**(4), 567–577 (1996)
24. Iordanou, H.N., Surgenor, B.W.: Experimental evaluation of the robustness of discrete sliding mode control versus linear quadratic control. IEEE Trans. Control Syst. Technol. **5**(2), 254–260 (1997)
25. Bartoszewicz, A.: Discrete-time quasi-sliding-mode control strategies. IEEE Trans. Industr. Electron. **45**(4), 633–637 (1998)
26. Hui, S., Żak, S.H.: On discrete-time variable structure sliding mode control. Syst. Control Lett. **38**(4), 283–288 (1999)
27. Golo, G., Milosavljević, Č.: Robust discrete-time chattering free sliding mode control. Syst. Control Lett. **41**(1), 19–28 (2000)
28. Koshkouei, A.J., Zinober, A.: Sliding mode control of discrete-time systems. J. Dyn. Syst. Meas. Contr. **122**, 793 (2000)
29. Utkin, V.: Variable structure systems with sliding modes. IEEE Trans. Autom. Control **22**(2), 212–222 (1977)
30. Ramírez, H.S., Battle, V.F.: A generalized pi sliding mode and pwm control of switched fractional systems. Modern sliding mode control theory, pp. 201–221. Springer, Berlin (2008)
31. Si-Ammour, A., Djennoune, S., Bettayeb, M.: A sliding mode control for linear fractional systems with input and state delays. Commun. Nonlinear Sci. Numer. Simul. **14**(5), 2310–2318 (2009)
32. Pisano, A., Rapaić, M., Usai, E.: Second-order sliding mode approaches to control and estimation for fractional order dynamics. Sliding modes after the first decade of the 21st century, pp. 169–197. Springer, Berlin (2012)
33. Kamal, S., Raman, A., Bandyopadhyay, B.: Discrete sliding mode control of fractional linear systems. In: IEEE international symposium on industrial electronics (ISIE), pp. 820–825. (2012)
34. Hilfer, R.: Appl. Fractional Calc. Phys. Word Scientific, Singapore (2000)
35. Ostalczyk, P.: The non-integer difference of the discrete-time function and its application to the control system synthesis. Int. J. Syst. Sci. **31**(12), 1551–1561 (2000)
36. Dzielinski, A., Sierociuk, D.: Ultracapacitor modelling and control using discrete fractional order state-space model. Acta Montanist. Slovaca **13**(1), 136–145 (2008)

Chapter 8
Disturbance Observer Based Robust Control for Fractional Order Systems

8.1 Introduction

Similar to integer order systems, fractional order systems have also two approaches to implement the controller in practical systems. In the first approach, continuous controller is designed for the continuous time fractional order system and then it is implemented real plant in discrete fashion. However, simply discretization of the continuous fractional order controller never achieves desired objective as the control is not applied in continuous manner, even it may result unstable behavior. In the second approach, design the discrete controller by discretizing the continuous fractional order system and then apply it at discrete instants. So, in this chapter, we have adopted this methodology to design discrete fractional order control.

To the best of author's knowledge robust control of discrete fractional order system still one of the open problems in literature. In Chap. 7 some new results on discrete time sliding mode control for fractional order system is proposed. But much more work needs to be done in the area of robust control of discrete fractional order system. Our current work proposes a novel and simple solution based on disturbance observer (DOB) to the addressed problem.

Controller design based on DOB is an effective method to achieve robustness against disturbances and model uncertainties. However, it is limited to integer order system only. The basic working principle of DOB is to provide an estimate of the actual disturbance, which is used to compensate the disturbances. The DOB is used only for disturbance rejection, however the overall performance of the closed loop system is dependent on the nominal control. Therefore, the implementation of control requires two loop in which the inner loop is just like asymptotic or finite time observer for estimating the disturbance and the outer loop of control contains the nominal control which is required to achieve the desired control performances.

In this chapter we propose a new methodology for the robust stabilization of discrete fractional order systems [1]. For achieving the specified goal a disturbance observer is constructed based on full state information. After that an extra compensator based on theory of disturbance observer is added in the classical state feedback.

B. Bandyopadhyay and S. Kamal, *Stabilization and Control of Fractional Order Systems: A Sliding Mode Approach*, Lecture Notes in Electrical Engineering 317,
DOI 10.1007/978-3-319-08621-7_8

It is also shown that after applying the control closed loop system would become robust against the disturbance as the estimation error gets small.

The chapter has been organized in the following way: In Sect. 8.2, some preliminaries of discrete fractional order system have been presented. The main result is presented in Sect. 8.3. In Sect. 8.4, state feedback control design for the discrete fractional order system is presented. Finally, some concluding remarks are included in Sect. 8.5.

8.2 Preliminaries of Discrete Fractional Order System

Consider a discrete fractional order linear system [2]

$$\begin{aligned} \Delta^{\alpha} x(k+1) &= Ax(k) + Bu(k) + Ed(k), \quad x(0) = x_0, \\ y(k) &= Cx(k), \end{aligned} \tag{8.1}$$

where $x \in \mathbb{R}^n$, $u \in \mathbb{R}^m$, $d \in \mathbb{R}^q$, $y \in \mathbb{R}^l$, $A \in \mathbb{R}^{n\times n}$, $B \in \mathbb{R}^{n\times m}$, $E \in \mathbb{R}^{n\times q}$ and $C \in \mathbb{R}^{n\times l}$ are the states, control input, disturbance, measurement output, system matrix, input matrix, disturbance matrix and output matrix respectively. In this model, order of the differentiation α is taken same for all state variable. The disturbance is assumed to be unknown but slowly time-varying and satisfying the following assumption:

Assumption 8.1 $d(k) = [d^1(k), d^2(k), \ldots, d^q(k)]^T$ is slowly time varying disturbances such that, for some constants ε_i

$$\left|\Delta d^i(k)\right| \le h\varepsilon_i, \quad \forall\, k \ge 1, (i = 1, \ldots, q) \tag{8.2}$$

where $\Delta d^i(k) = d^i(k) - d^i(k-1)$ and h is the sampling period.
Using the definition given by Grunwald, one can also write

$$x(k+1) = h^{\alpha}\Delta^{\alpha} x(k+1) - \sum_{j=1}^{k+1} (-1)^j \binom{\alpha}{j} x(k-j+1). \tag{8.3}$$

Substituting (8.3) into (8.1)

$$\begin{aligned} x(k+1) = \tilde{A}x(k) - \sum_{j=1}^{k+1} (-1)^j \binom{\alpha}{j} x(k-j+1) \\ + \tilde{B}u(k) + \tilde{E}d(k), \end{aligned} \tag{8.4}$$

where $\tilde{A} = h^{\alpha} A$, $\tilde{B} = h^{\alpha} B$ and $\tilde{E} = h^{\alpha} E$.

The above description can be extended to discrete fractional order system with different (non-commensurate) orders

$$\Delta_h^{\gamma} x(k+1) = Ax(k) + Bu(k) + Ed(k), \quad x(0) = x_0$$
$$x(k+1) = \Delta_h^{\gamma} x(k+1) + \sum_{j=1}^{k+1} A_j x(k-j+1), \tag{8.5}$$

where,

$$\Delta_h^{\gamma} x(k+1) = \begin{bmatrix} h^{\gamma_1} \Delta^{\gamma_1} x_1(k+1) \\ \vdots \\ h^{\gamma_n} \Delta^{\gamma_n} x_n(k+1) \end{bmatrix}, \tag{8.6}$$

where $\gamma_i, i = 1, 2, \ldots, n$ denotes the fractional orders. One can write, the following state space model from (8.5) and (8.6)

$$x(k+1) = \sum_{j=0}^{k} A_j x(k-j) + \tilde{B}u(k) + \tilde{E}d(k), \tag{8.7}$$

where

$$A_0 = \tilde{A} + \text{diag}\binom{\gamma_i}{1}, \quad i = 1, 2, \ldots, n$$
$$A_j = \text{diag}\left(-(-1)^{j+1}\binom{\gamma_i}{j+1}\right), \quad j = 1, 2, \ldots, n. \tag{8.8}$$

and $\tilde{A} = [\tilde{A}_1, \tilde{A}_2, \ldots, \tilde{A}_n]^{\top}$, $\tilde{B} = [\tilde{B}_1, \tilde{B}_2, \ldots, \tilde{B}_n]^{\top}$, $\tilde{E} = [\tilde{E}_1, \tilde{E}_2, \ldots, \tilde{E}_n]^{\top}$, $\tilde{A}_i = h^{\gamma_i} A_i$, $\tilde{B}_i = h^{\gamma_i} B_i$, $\tilde{E}_i = h^{\gamma_i} E_i$ and A_i, B_i, E_i denotes the rows of A, B and E respectively.

Remark 8.2 Before finding the stabilizing control for the uncertain system (8.7), we proposed the disturbance observer for estimating the disturbance in the next section.

8.3 Design of the Disturbance Observer in Presence of Full State Information

Theorem 8.3 *Disturbance observer for the uncertain system* (8.7) *be proposed as*

$$\hat{d}(k) = Lx(k) - z(k)$$
$$z(k+1) = z(k) + L\left\{\sum_{j=0}^{k} A_j x(k-j) - x(k) + \tilde{B}u(k)\right\} + L\tilde{E}\hat{d}(k), \tag{8.9}$$

where $L \in \mathbb{R}^{q\times n}$, $\hat{d}(k) \in \mathbb{R}^q$ is the disturbance estimation, $z(k) \in \mathbb{R}^q$ is the fictitious variable. Then, the state estimation error, $e(k) = d(k) - \hat{d}(k)$, has the dynamics

$$e(k+1) = \left(I_q - L\tilde{E}\right) e(k) + \Delta d(k+1), \tag{8.10}$$

where $\Delta d(k+1) = d(k+1) - d(k)$.

Proof Expression of error dynamics is formulated as

$$e(k+1) = d(k+1) - \hat{d}(k+1). \tag{8.11}$$

Substituting $\hat{d}(k+1)$ from (8.9)

$$\begin{aligned} e(k+1) =& d(k+1) - (Lx(k+1) - z(k+1)) \\ =& d(k+1) - L\left(\sum_{j=0}^{k} A_j x(k-j) + \tilde{B}u(k) + \tilde{E}d(k)\right) \\ &+ z(k) + L\left\{\sum_{j=0}^{k} A_j x(k-j) - x(k) + \tilde{B}u(k)\right\} + L\tilde{E}\hat{d}(k) \\ =& d(k+1) - \left(Lx(k) - z(k)) - L\tilde{E}(d(k) - \hat{d}(k)\right) \\ =& d(k+1) - d(k) + d(k) - (Lx(k) - z(k)) - L\tilde{E}\left(d(k) - \hat{d}(k)\right) \\ =& \Delta d(k+1) + \left(I_q - L\tilde{E}\right) e(k). \end{aligned} \tag{8.12}$$

Remark 8.4 Stability of the proposed disturbance observer can be ensured if the pair $(I_q, \tilde{E})$ is observable. It is not hard to prove that pair $(I_q, \tilde{E})$ is observable, since $\text{rank}(\tilde{E}) = q$. Hence, it is obvious that we will be always able to estimate the slow varying disturbance within a bound when the full state information is available.

There are infinite possible choices for the L, such that error dynamics will be stable. One of the simplest choice is stated here

Lemma 8.5 *A given matrix N, which is satisfying $N\tilde{E}$ left-invertible* (i.e., $(N\tilde{E})^+ N\tilde{E} = I_q$), suppose that $L = (I_q - \Omega)(N\tilde{E})^+ N$ where $\Omega = \text{diag}\left\{\lambda_1, \ldots, \lambda_q\right\}$ with $|\lambda_i| < 1 (i = 1, \ldots, q)$. This results in each error dynamics as follows:

$$e^i(k+1) = \lambda_i e^i(k) + \Delta d^i(k+1), \tag{8.13}$$

where $\Delta d^i(k+1) = d^i(k+1) - d^i(k)$. Then, the estimation error confined within a bound ε^q

$$\varepsilon^q = \sqrt{\sum_{i=1}^{q} \left(\frac{1}{1-\lambda_i} h\varepsilon_i \right)^2}. \tag{8.14}$$

Proof If we start from $e^i(0)$, where $i = 1, 2, \ldots, q$ the future values of $e^i(k)$ can be written as

$$e^i(k) = \lambda_i^k e^i(0) + \sum_{i=0}^{k-1} \lambda_i^k \Delta d^i(k+1). \tag{8.15}$$

If $\Delta d^i(k+1) < h\varepsilon_i$, then the above Eq. (8.15) can be written as

$$\left| e^i(k) \right| < \left| \lambda_i^k e^i(0) \right| + \sum_{i=0}^{k-1} \lambda_i^k h\varepsilon_i. \tag{8.16}$$

For sufficiently large k the right hand side of the above inequality approaches to $\varepsilon_i = \frac{1}{1-\lambda_i} h\varepsilon_i$. For each disturbance vector similar relation is obtained after taking large value of k. One can calculate the average value of band by taking Euclidean norm of all ε, which is same as (8.14).

8.4 State Feedback Control Design for the Discrete Fractional Order System

8.4.1 Classical State Feedback Design for the Disturbance Free Discrete Fractional Order System

Classical state-feedback control for the system (8.7) ($d(k) = 0$) is given as [3]

$$u(k) = \sum_{j=0}^{k} \Xi_j x(k-j). \tag{8.17}$$

Closed loop system after substituting control $u(k)$ into (8.7)

$$\begin{aligned} x(k+1) &= \sum_{j=0}^{k} \left(A_j + \tilde{B}\Xi_j \right) x(k-j), \\ &= \sum_{j=0}^{k} A_{\Xi j} x(k-j). \end{aligned} \tag{8.18}$$

where $A_{\Xi j} = (A_j + \tilde{B}\Xi_j)$. One can also write

$$
\begin{aligned}
x(1) &= \left(A_0 + \tilde{B}\Xi_0\right)x(0) \\
x(2) &= \left(A_0 + \tilde{B}\Xi_0\right)x(1) + \left(A_1 + \tilde{B}\Xi_1\right)x(0) \\
&\vdots \\
x(k) &= \left(A_0 + \tilde{B}\Xi_0\right)x(k-1) + \cdots + \left(A_{k-1} + \tilde{B}\Xi_{k-1}\right)x(0) \\
&\vdots
\end{aligned} \tag{8.19}
$$

Further in matrix form one can write

$$
\begin{bmatrix} x(1) \\ x(2) \\ \vdots \\ x(k) \\ \vdots \end{bmatrix} = \begin{bmatrix} (A_0 + \tilde{B}\Xi_0) & 0 & 0 & 0 & \cdots \\ (A_1 + \tilde{B}\Xi_1) & (A_0 + \tilde{B}\Xi_0) & 0 & 0 & \cdots \\ \vdots & \vdots & \vdots & \vdots & \vdots \\ (A_{k-1} + \tilde{B}\Xi_{k-1}) & \cdots & \cdots & \cdots & \cdots \\ \vdots & \vdots & \vdots & \vdots & \vdots \end{bmatrix} \begin{bmatrix} x(0) \\ x(1) \\ \vdots \\ x(k-1) \\ \vdots \end{bmatrix}. \tag{8.20}
$$

Considering an infinite column sequence $\tilde{x}$, and using the backward shift $\tilde{S}$, Eq. (8.20), can be written as

$$
\tilde{x} = \tilde{S}\tilde{A}_{\Xi}\tilde{x} \tag{8.21}
$$

where

$$
\tilde{A}_{\Xi} = \begin{bmatrix} (A_0 + \tilde{B}\Xi_0) & 0 & 0 & 0 & \cdots \\ (A_1 + \tilde{B}\Xi_1) & (A_0 + \tilde{B}\Xi_0) & 0 & 0 & \cdots \\ \vdots & \vdots & \vdots & \vdots & \vdots \\ (A_{k-1} + \tilde{B}\Xi_{k-1}) & \cdots & \cdots & \cdots & \cdots \\ \vdots & \vdots & \vdots & \vdots & \vdots \end{bmatrix}, \quad \tilde{S} = \begin{bmatrix} 0 & 0 & 0 & 0 & \cdots \\ I_n & 0 & 0 & 0 & \cdots \\ \vdots & \vdots & \vdots & \vdots & \vdots \\ 0 & \cdots & \cdots & I_n & \cdots \\ \vdots & \vdots & \vdots & \vdots & \vdots \end{bmatrix}, \tag{8.22}
$$

$\tilde{A}_{\Xi}$ is further decomposed as

$$
\begin{aligned}
\tilde{A}_{\Xi} &= \begin{bmatrix} A_0 & 0 & 0 & 0 & \cdots \\ A_1 & A_0 & 0 & 0 & \cdots \\ \vdots & \vdots & \vdots & \vdots & \vdots \\ A_{k-1} & \cdots & \cdots & \cdots & \cdots \\ \vdots & \vdots & \vdots & \vdots & \vdots \end{bmatrix} + \begin{bmatrix} \tilde{B} & 0 & 0 & 0 & \cdots \\ 0 & \tilde{B} & 0 & 0 & \cdots \\ \vdots & \vdots & \vdots & \vdots & \vdots \\ \cdots & \cdots & \cdots & \cdots & \cdots \\ \vdots & \vdots & \vdots & \vdots & \vdots \end{bmatrix} \begin{bmatrix} \Xi_0 & 0 & 0 & 0 & \cdots \\ \Xi_1 & \Xi_0 & 0 & 0 & \cdots \\ \vdots & \vdots & \vdots & \vdots & \vdots \\ \cdots & \cdots & \cdots & \cdots & \cdots \\ \vdots & \vdots & \vdots & \vdots & \vdots \end{bmatrix} \\
&= \tilde{A} + \tilde{B}\tilde{\Xi}.
\end{aligned} \tag{8.23}
$$

Equation (8.21) is further written as

$$\tilde{x} = \tilde{S}\tilde{A}_{\Xi}\tilde{x} = \tilde{S}\tilde{A}\tilde{x} + \tilde{S}\tilde{B}\tilde{\Xi}\tilde{x}, \tag{8.24}$$

where

$$\tilde{\Xi} = \begin{bmatrix} \Xi_0 & 0 & 0 & 0 & \cdots \\ \Xi_1 & \Xi_0 & 0 & 0 & \cdots \\ \Xi_2 & \Xi_1 & \Xi_0 & 0 & 0 \\ \cdots & \cdots & \cdots & \cdots & \cdots \\ \vdots & \vdots & \vdots & \vdots & \vdots \end{bmatrix}. \tag{8.25}$$

The reachability of discrete fractional order system (8.25) under the operator-theoretic framework is stated as

Definition 8.6 [3] The pair $(\tilde{A}, \tilde{B})$ is K-step reachable if the operator reachability Gramian defined by

$$\tilde{W}_r^p = \tilde{C}_p^T \tilde{C}_p, \tag{8.26}$$

where

$$\tilde{C}_p = \begin{bmatrix} \tilde{B} & \tilde{A}\tilde{B} & \cdots & \tilde{A}^{p-1}\tilde{B} \end{bmatrix}, \tag{8.27}$$

is invertible.

Let us now define the transition matrix

$$G_{\Xi k} = \begin{cases} I_n & \text{for } k = 0 \\ \sum_{j=0}^{k-1} A_{\Xi j} G_{\Xi (k-1-j)} & \text{for } k \geq 1 \end{cases} \tag{8.28}$$

Now our aim is to choose the state feedback operator gain $\tilde{\Xi}$, such that system (8.24) is stable and also ensures

$$\|G_{\Xi k}\| \leq 1 \ \text{ for } k \geq 1. \tag{8.29}$$

There are infinite possible choice for the $\tilde{\Xi}$, one of the easier ways for choosing $\tilde{\Xi}$ is stated as following:

Lemma 8.7 [3] *Suppose that $(\tilde{A}, \tilde{B})$ is k-reachable and let*

$$\tilde{\Xi} = -\tilde{B}^T \tilde{A}^T (\tilde{w}_r^p)^{-1} \tilde{A}^{p+1}, \tag{8.30}$$

then the closed-loop system (8.24) *is asymptotically stable, that is* $\rho(\tilde{A}_{\Xi S}) < 1$, *where* $\tilde{A}_{\Xi S} = \tilde{S}\tilde{A}_{\Xi}$.

Remark 8.8 Above controller gives satisfactory performance only, if the system is disturbance free. But in the real seniors actually system is never disturbance free. For making classical controller robust in the presence of disturbance, we add an extra compensator to the existing nominal controller based on theory of disturbance observer which is already proposed in the above section. In the next subsection we present the controller for the uncertain discrete fractional order system.

8.4.2 Proposed State Feedback Design for the Uncertain Discrete Fractional Order System

Proposed control input for the system (8.7) contains two part

- Nominal controller is derived from the classical theory when discrete fractional order system is free from disturbance. Expression for the nominal controller is

$$
\begin{aligned}
u_{\text{nominal}}(k) &= \sum_{j=0}^{k} \Xi_j x(k-j) \\
&= -\sum_{j=0}^{k} \tilde{B}^T \tilde{A}^T \left(\tilde{w}_r^p\right)^{-1} \tilde{A}^{p+1} x(k-j).
\end{aligned} \tag{8.31}
$$

- Second part of control is based on disturbance observer theory, also known as compensator. Expression for the compensator is

$$
u_{\text{D}}(k) = -\tilde{B}^{+}\tilde{E}\hat{d}(k). \tag{8.32}
$$

Hence overall controller for the system (8.7) is given as

$$
u(k) = u_{\text{nominal}}(k) + u_{\text{D}}(k). \tag{8.33}
$$

Closed loop dynamics after applying the control (8.33) becomes

$$
\begin{aligned}
x(k+1) = &\sum_{j=0}^{k} A_j x(k-j) + \tilde{B} u_{\text{nominal}}(k), \\
&+ (I_q - \tilde{B}\tilde{B}^{+})\tilde{E}d(k) + \tilde{B}\tilde{B}^{+}\tilde{E}e(k).
\end{aligned} \tag{8.34}
$$

Now suppose the particular case of matched disturbance (i.e., $u(k)$ and $d(k)$ both belongs to $\mathbb{R}^m$). Hence $(I_q - \tilde{B}\tilde{B}^{+})\tilde{E} = 0$ and $\tilde{B}\tilde{B}^{+}\tilde{E} = \tilde{E}$.

$$x(k+1) = \sum_{j=0}^{k} A_j x(k-j) + \tilde{B} u_{\text{nominal}}(k) + \tilde{E} e(k). \tag{8.35}$$

It is clear that the closed loop dynamics become robust against the disturbance because the estimation error gets small which is discussed in Sect. 8.3.

8.5 Summary

Robust control of discrete fractional order system is one of the open problems in literature for the theoretical and practical point of view. In this Chapter we are trying to give one of the possible solutions of the specified problem. For achieving the robust state feedback a disturbance observer is constructed based on full state information. After that an extra compensator based on the theory of disturbance observer is added to the classical state feedback.

References

1. Kamal, S., Bandyopadhyay, B.: Robust controller design for discrete fractional order system: A disturbance observer based approach. In: Advances in Control and Optimization of Dynamical Systems, vol. 3, pp. 558–563 (2014)
2. Raynaud, H.F., Zergaınoh, A.: State-space representation for fractional order controllers. Automatica **36**(7), 1017–1021 (2000)
3. Guermah, S., Djennoune, S., Bettayeb, M.: Discrete-time fractional-order systems: Modeling and stability issues (2012)

Chapter 9
Contraction Analysis by Integer Order and Fractional Order Infinitesimal Variations

9.1 Introduction

There are several types of stability concepts proposed in control theory depending on how and when the system is stabilized. One of the most classic concepts regarding stability is obtained by introducing a weighted norm, or more generally by defining a Lyapunov function, which has to decrease at each instant and is strongly related to the norm of the system states. The revisiting concept of contraction leads to the introduction of suitable Riemann metrics or more generally Finsler metrics. Recently, inspired from fluid mechanics and differential geometry, Lohmiller and Slotine proposed a new method of stability analysis known as *contraction theory* ([1–3]). This theory is based on the concept that the stability can be analyzed differentially by analyzing if the nearby trajectories converge to one another, rather than through finding some implicit motion integral as in Lyapunov theory, or through some global state transformation as in feedback linearization (see [4]).

Making use of the concepts of contraction theory we have attempted to design a globally exponentially stable controller for fractional order systems [5]. To the best of the author's knowledge this is the first attempt in this direction.

Classical contraction theory which is proposed in literature [1, 2], able to analyze the convergence behavior of nonlinear systems in state space only when the system is continuously differentiable. But, lot of systems existing in real time applications where the systems are modeled using non differentiable equations. This chapter, revisited the contraction theory by replacing the first order variation of system state by the fractional order variation [6, 7]. The main advantage of the proposed approach is that it also works for analyzing the stability of non differentiable systems. Also, this approach is useful for analyzing stability of fractional order systems and designing the fractional order controller possibly better than the integer order controller, for dynamical systems described by integer order differential equation [8]. To the best of the author's knowledge this is the first attempt in this direction.

The Chapter has been organized in the following way: In Sect. 9.2 a brief review of fractional order Routh-Hurwitz condition has been presented. Section 9.3 introduces

B. Bandyopadhyay and S. Kamal, *Stabilization and Control of Fractional Order Systems: A Sliding Mode Approach*, Lecture Notes in Electrical Engineering 317,
DOI 10.1007/978-3-319-08621-7_9

the concept of contraction theory. The motivation behind the present work is given in Sect. 9.4. The main result is presented in Sect. 9.5. In Sect. 9.6, a numerical example is presented to illustrate the proposed method. The contraction analysis using fractional order infinitesimal variation is presented in Sect. 9.7. In Sect. 9.8, same approach is extended for checking the contraction behavior of fractional order systems. Finally, some concluding remarks are included in Sect. 9.9.

9.2 A Brief Review of Fractional Order Routh-Hurwitz Conditions

Consider the system $D^{\alpha}x = Ax$. The problem of interest is that all the roots of the polynomial equation [9]

$$P(\lambda) = 0, \quad P(\lambda) = \lambda^n + a_1\lambda^{n-1} + a_2\lambda^{n-2} + \cdots + a_n, \tag{9.1}$$

satisfy

$$|arg(\lambda)| > \alpha\pi/2, \tag{9.2}$$

where all the coefficients in (9.1) are real.

For integer order ($\alpha = 1$) the solution yields the Routh-Hurwitz conditions

$$0 < a_1, \quad 0 < \begin{pmatrix} a_1 & 1 \\ a_3 & a_2 \end{pmatrix}, \quad 0 < \begin{pmatrix} a_1 & 1 & 0 \\ a_3 & a_2 & a_1 \\ a_5 & a_4 & a_3 \end{pmatrix} \cdots \tag{9.3}$$

For $\alpha \in [0, 1)$ these conditions are sufficient but not necessary.

Definition 9.1 The discriminant $D(f)$ of a polynomial

$$f(x) = x^n + a_1x^{n-1} + a_2x^{n-2} + \cdots + a_n,$$

is defined by $D(f) = (-1)^{n(n-1)/2}R(f, f')a_n^{n-k-2}$ where f' is the derivative of f and k is the degree of f' and $R(f, f')$ equal to the determinant of the $(2n-1)\times(2n-1)$ Sylvester matrix whose $n-1$ first rows contain the coefficients of f and the n last ones the coefficients of its derivative.

Proposition 9.2 *1. For $n = 1$ the condition for (9.2) is $a_1 > 0$.*

2. *For $n = 2$ the conditions for (9.2) are either the Routh-Hurwitz conditions or $a_1 < 0, 4a_2 > (a_1)^2, \left|tan^{-1}(\sqrt{4a_2 - (a_1)^2})/a_1\right| > \alpha\pi/2$.*
3. *For $n = 3$ if the discriminant of $p(\lambda)$, $D(P)$ is positive then the Routh-Hurwitz conditions are the necessary and sufficient conditions for (9.2), i.e.,*

$$a_1 > 0, \quad a_3 > 0 \;\; a_1a_2 > a_3 \;\; if \;\; D(P) > 0. \tag{9.4}$$

4. *If* $D(P) < 0$, $a_1 \geq 0$, $a_2 \geq 0$, $a_3 > 0$, $\alpha < (2/3)$ *then (9.2) is satisfied. Also if* $D(P) < 0$, $a_1 < 0$, $a_2 < 0$, $\alpha > (2/3)$ *then all roots of* $P(\lambda)$ *satisfy* $|arg(\lambda)| < \alpha\pi/2$.
5. *If* $D(P) < 0$, $a_1 > 0$, $a_2 > 0$, $a_1 a_2 = a_3$ *then (9.2) is satisfied for all* $\alpha \in [0, 1)$.
6. *For general n,* $a_n > 0$, *is a necessary condition for (9.2).*
7. *If* $\forall\lambda$, $P(\lambda) = P(-\lambda)$ *then define* $x = \lambda^2$ *and the Routh-Hurwitz conditions for the resulting polynomial in x are necessary conditions for (9.2) for all* $\alpha \in [0, 1)$.
8. *For* $n > 1$, *the necessary and sufficient condition for (9.2) is*

$$\int_0^{\infty} dz/P(z)|_{C2} + \int_{-\infty}^{0} dz/P(z)|_{C1} = 0, \tag{9.5}$$

where $C1$ *is the curve* $z = x(1 - i tan\alpha\pi/2)$, *and* $C2$ *is the curve* $z = x(1 - i tan\alpha\pi/2)$, $i = \sqrt{-1}$

Note Using the results of [10], if $D(f) > 0$ (< 0) then there is an even (odd) number of pairs of complex roots for the equation $f(x) = 0$. For $n = 3$ this implies that $D(f) > 0$ implies that all the roots are real and $D(f) < 0$ implies that there is only one real root and one complex root and its complex conjugate. For $n = 3$ we have

$$D(P) = 18a_1a_2a_3 + (a_1a_2)^2 - 4a_3(a_1)^3 - 4(a_2)^3 - 27(a_3)^2. \tag{9.6}$$

Proof See [9] □

Lemma 9.3 *[9] For all* $n > 3$ *if* $\Delta_1, \Delta_2, \ldots, \Delta_n$ *are Routh Hurwitz determinants*

$$\Delta_1 = a_1, \quad \Delta_2 = \begin{vmatrix} a_1 & 1 \\ a_3 & a_2 \end{vmatrix}, \quad \Delta_3 = \begin{vmatrix} a_1 & 1 & 0 \\ a_3 & a_2 & a_1 \\ a_5 & a_4 & a_3 \end{vmatrix}$$

then the conditions $\Delta_i > 0$ *for* $i = 1, 2, \ldots, n-2$, $a_n > 0$, $\Delta_{n-1} = 0$, *are sufficient conditions that* (9.1) *is valid for all* $\alpha \in [0, 1)$.

Fractional-order integration and differentiation are the generalization of their integer-order counterparts. Efforts to extend the specific definition of the traditional integer-order to the more general arbitrary order produced different definitions for fractional derivatives. One of the most common definitions used is the Reimann-Liouville definition (see [11]). RL definition is considered henceforth unless otherwise specified. In the next section we present a brief review of contraction analysis of the dynamical system, which is essential for proving the the main result of this chapter.

9.3 Contraction Analysis of Dynamical Systems

Consider an autonomous dynamical system [2]

$$\dot{x} = f(x(t)), \tag{9.7}$$

where f is a nonlinear vector field and $x(t)$ is an n-dimensional state vector. It is assumed that all the quantities are real and smooth and thus all the required derivatives exist and are continuous. In contraction analysis, the state x is slightly modified to see the change in velocity vector $\dot{x}$, that is termed as the virtual displacement (infinitesimal displacement at fixed time) δx and virtual velocity $\delta \dot{x}$ respectively, introduced by Lagrange. Due to this infinitesimal displacement of the state, virtual dynamics are introduced into the system, which can be represented by,

$$\delta \dot{x} = \delta f(x(t)) = \frac{\partial f(x(t))}{\partial x} \delta x. \tag{9.8}$$

A locally state dependent nonsingular transformation Θ

$$\delta z = \Theta \delta x(t), \tag{9.9}$$

transforms the virtual dynamics (9.8) into δz-coordinates as

$$\frac{d}{dt} \delta z = \dot{\Theta} \delta x(t) + \Theta \delta \dot{x}(t) = F \delta z, \tag{9.10}$$

where F is termed as generalized Jacobian, given by

$$F = \left(\dot{\Theta} + \Theta \frac{\partial f}{\partial x} \right) \Theta^{-1}. \tag{9.11}$$

The rate of change of squared length is given by

$$\frac{d}{dt} (\delta z^T \delta z) = 2 \delta z^T F \delta z, \tag{9.12}$$

A more general definition of infinitesimal length can be given by

$$\delta x^T(t) M(x(t)) \delta x(t), \tag{9.13}$$

where $M(x(t))$ is a symmetric, uniformly positive definite and continuously differentiable metric (formally, this defines a Riemannian manifold). Using the more general definition of infinitesimal length, one can calculate its rate of change as

$$
\begin{aligned}
&\frac{d}{dt}(\delta x^T(t)M(x(t))\delta x(t)) \\
&= \delta x^T(t)\left(\frac{\partial f^T}{\partial x}M + M\frac{\partial f}{\partial x} + \dot{M}\right)\delta x(t),
\end{aligned}
\tag{9.14}
$$

Based on the above observation, [1] gives the following definition of the contraction region as:

Definition 9.4 Given the n-dimensional system equations $\dot{x} = f(x(t))$, a region of the state space is called a contraction region with respect to a uniformly positive definite metric $M(x(t)) = \Theta^T\Theta$, if equivalently F in Eq. (9.11) or $\frac{\partial f^T}{\partial x}M + M\frac{\partial f}{\partial x} + \dot{M}$ are uniformly negative definite in that region.

Using the above definition, [1] generalized the convergence result as:

Theorem 9.5 *Given the system equations $\dot{x} = f(x(t))$, any trajectory, which starts in a ball of constant radius with respect to the metric $M(x(t))$, centered at a given trajectory and contained at all times in a contraction with respect to $M(x(t))$, remains in that ball and converges exponentially to this trajectory.*

Proof See Lohmiller and Slotine (1998). □

A similar theorem (see [2]) shows that contraction metrics can be used to prove convergence to a single trajectory, and thus existence and/or uniqueness of equilibria.

Theorem 9.6 *Consider the autonomous system $\dot{x} = f(x(t))$. If a contraction metric exists for the system over the entire state-space and a finite equilibrium exists, then this equilibrium is unique and all trajectories converge to this equilibrium. If the system is exponentially contracting, there exists a unique finite equilibrium, and all trajectories converge to this equilibrium.*

In terms of the Jacobian matrix, [1] gives the following definition for finding a contraction region:

Definition 9.7 Given the system equations $\dot{x} = f(x, t)$, a region of the state space is called a contraction or semi-contraction region, if the Jacobian matrix $\partial f/\partial x$ is uniformly negative definite or negative semi-definite in that region.

The main result of this Chapter is inspired from the finite time stabilization of an integrator chain [12], which is presented in next section.

9.4 Motivation: Finite Time Stabilization of an Integrator Chain

Consider the nominal system (9.15), which is represented by SISO independent integrator chains, defined as follows

$$\begin{cases} \dot{z}_1 & = z_2 \\ & \vdots \\ \dot{z}_{n-1} & = z_n \\ \dot{z}_n & = u. \end{cases} \tag{9.15}$$

Theorem 9.8 *[12] Let $k_1, \ldots, k_n > 0$ be such that the polynomial $\lambda^n + k_n\lambda^{n-1} + \cdots + k_2\lambda + k_1$ is Hurwitz. Consider system (9.15). There exists $\varepsilon \in (0, 1)$ such that, for every $\alpha \in (1-\varepsilon, 1)$, the origin is a globally finite time stable equilibrium for the system under the feedback*

$$u(z) = -k_1 sgn(z_1)|z_1|^{\alpha_1} - \cdots - k_n sgn(z_n)|z_n|^{\alpha_n}, \tag{9.16}$$

where $\alpha_1, \ldots, \alpha_n$ satisfy

$$\alpha_{i-1} = \frac{\alpha_i \alpha_{i+1}}{2\alpha_{i+1} - \alpha_1}, \quad i = 2, \ldots, n, \quad \text{with} \quad \alpha_{n+1} = 1. \tag{9.17}$$

Remark 9.9 Above result is only limited for the integer order systems. But in the last two decades, fractional chain of integrators have been used to model various stable physical phenomena. Therefore, in next section we generalize this result for the fractional order system.

9.5 Contraction Analysis of Fractional Order System by Integer Order Infinitesimal Variation

Consider the following fractional order system:

$$\begin{cases} D^\alpha x_1 & = x_2 \\ D^\alpha x_2 & = x_3 \\ & \vdots \\ D^\alpha x_{n-1} & = x_n \\ D^\alpha x_n & = u. \end{cases} \tag{9.18}$$

This can be also represented as:

$$D^{\alpha}x = Ax + Bu,$$

where $D^{\alpha}x = [D^{\alpha}x_1\ D^{\alpha}x_2 \cdots D^{\alpha}x_n]^T = D^{\alpha}[x_1\ x_2 \cdots x_n]^T$,

$$A = \begin{pmatrix} 0 & 1 & \cdots & 0 \\ \vdots & \vdots & \ddots & \vdots \\ 0 & 0 & \vdots & 1 \\ 0 & 0 & 0 & 0 \end{pmatrix} \quad \text{and} \quad B^T = \begin{pmatrix} 0 & 0 & \cdots & 1 \end{pmatrix} \tag{9.19}$$

and $x(t) \in \mathbb{R}^n$ represents the state vector, $u \in \mathbb{R}$ and $\alpha \in (0, 1)$. A and B are system matrices of appropriate dimensions. Note that (A, B) pair is controllable.

Remark 9.10 ([13]) For the commensurate case the above state space model remains valid with $D^{\alpha}x = D^{\alpha}[x_1\ x_2 \cdots x_n]^T$.

Following Theorem presents the main result of this Chapter, stabilization of fractional chain of integrators.

Theorem 9.11 *Let $k_1, \ldots, k_n > 0$ be such that the polynomial $\lambda^n + k_n\lambda^{n-1} + \cdots + k_2\lambda + k_1$ is Hurwitz (in the sense of fractional order systems. Consider the system (9.18). There exists $\varepsilon \in (0, 1)$ such that, for every $\beta_i \in (1 - \varepsilon, 1)$, where i=1, 2,... n, the origin is a globally exponential stable equilibrium for the system under the feedback:*

$$u = \left[\sum_{i=1}^{n} -k_i|x_i|^{\beta_i} sgn(x_i) - \sum_{j=1}^{n} \frac{(x_j)t^{\alpha-1}}{\Gamma(\alpha)} \right] \Omega, \tag{9.20}$$

where $\Omega = (t - \tau)^{(1-\alpha)}$. Here, τ is interpreted as the step size for numerical control implementation. The β_i's satisfy

$$\beta_{i-1} = \frac{\beta_i \beta_{i+1}}{2\beta_{i+1} - \beta_1}, \quad i = 2, \cdots, n, \quad \text{with} \quad \beta_{n+1} = 1. \tag{9.21}$$

Proof Consider,

$$D^{\alpha}x = Ax + Bu \tag{9.22}$$

Applying $D^{1-\alpha}$ to both side of (9.22), one can write

$$D^{1-\alpha}D^{\alpha}x = D^{1-\alpha}Ax + D^{1-\alpha}Bu,$$
$$\dot{x} = D^{1-\alpha}Ax + D^{1-\alpha}Bu.$$

Consider now two neighboring trajectories in the flow field of the above equation and the virtual displacement δx between them. This yields the following

$$\delta \dot{x} = \left[\frac{\partial D^{1-\alpha} Ax}{\partial x} + \frac{\partial D^{1-\alpha} Bu}{\partial x} \right] \delta x.$$

The squared distance between these two trajectories can be defined as $\delta x^T \delta x$, the rate of change of which is given by

$$\begin{aligned} \frac{d\delta x^T \delta x}{dt} &= 2\delta x^T \delta \dot{x} \\ &= 2\delta x^T \left[\frac{\partial D^{1-\alpha} Ax}{\partial x} + \frac{\partial D^{1-\alpha} Bu}{\partial x} \right] \delta x. \end{aligned}$$

Consider,

$$D^{1-\alpha} Ax = \frac{1}{\Gamma(1-(1-\alpha))} \frac{d}{dt} \int_0^t \frac{Ax(\tau)}{(t-\tau)^{((1-\alpha)-1+1)}} d\tau$$

$$\begin{aligned} \frac{\partial D^{1-\alpha} Ax}{\partial x} &= \frac{1}{\Gamma(\alpha))} \frac{d}{dt} \int_0^t \frac{A}{(t-\tau)^{(1-\alpha)}} d\tau \\ &= \frac{At^{\alpha-1}}{\Gamma(\alpha)}. \end{aligned}$$

For the proposed controller,

$$u = \left[\sum_{i=1}^{n} -k_i |x|^{\beta_i} sgn(x) - \sum_{j=1}^{n-1} \frac{(x_{j+1})t^{\alpha-1}}{\Gamma(\alpha)} \right] \Omega,$$

where $\Omega = (t-\tau)^{(1-\alpha)}$.

$$D^{1-\alpha} Bu = \frac{1}{\Gamma(1-(1-\alpha))} \frac{d}{dt} \int_0^t \frac{Bu(\tau)}{(t-\tau)^{((1-\alpha)-1+1)}} d\tau.$$

Substituting,

$$D^{1-\alpha} Bu = \frac{1}{\Gamma(1-(1-\alpha))} \frac{d}{dt} \int_0^t Bp(x) d\tau,$$

where

$$p(x) = \left[\sum_{i=1}^{n} -k_i |x_i|^{\beta_i} sgn(x_i) - \sum_{j=1}^{n} \frac{(x_j)t^{\alpha-1}}{\Gamma(\alpha)} \right].$$

$$
\begin{aligned}
D^{1-\alpha}Bu &= \frac{1}{\Gamma(\alpha))}\frac{d}{dt}\int_0^t p(x)d\tau,\\
\frac{\partial D^{1-\alpha}Bu}{\partial x} &= \frac{1}{\Gamma(\alpha)}\frac{d}{dt}\int_0^t \frac{\partial Bp(x)}{\partial x}d\tau\\
&= \frac{1}{\Gamma(\alpha)}\frac{\partial Bp(x)}{\partial x}\\
&= \frac{BP(x)}{\Gamma(\alpha)}.
\end{aligned}
$$

where,

$$
P(x) = \begin{pmatrix} 0 & 0 \cdots & 0 \\ \vdots & 0 \ddots & \vdots \\ 0 & 0 \ \vdots & 0 \\ -k_1|x_1|^{(\beta_1-1)} - \frac{t^{\alpha-1}}{\Gamma(\alpha)} & \cdots\cdots & -k_n|x_n|^{(\beta_n-1)} - \frac{t^{\alpha-1}}{\Gamma(\alpha)} \end{pmatrix} \tag{9.23}
$$

The Jacobian is represented by:

$$
\begin{aligned}
J &= \left[\frac{\partial D^{1-\alpha}Ax}{\partial x} + \frac{\partial D^{1-\alpha}Bu}{\partial x}\right]\\
&= \frac{At^{\alpha-1}}{\Gamma(\alpha)} + \frac{BP(x)}{\Gamma(\alpha)}
\end{aligned} \tag{9.24}
$$

One can easily observe that as $t \to \infty$, the symmetric part of the Jacobian matrix J becomes

$$
J = \begin{pmatrix} 0 & \to 0 & \cdots & 0 \\ \vdots & 0 & \ddots & \vdots \\ 0 & 0 & \vdots & 0 \\ (-)\text{ve} & \cdots & (-)\text{ve} & (-)\text{ve} \end{pmatrix}. \tag{9.25}
$$

which is negative semi-definite, only when $k_i > 0, i = 1, 2, \ldots, n$. Hence, the rate of change of the distance between the two considered trajectories is negative. As a result, the distance progressively decreases. By path integration, this implies that the length of any finite path converges exponentially to zero.

More discussion is required for the selection of k_i, because of the different scaling property satisfied by fractional order integral when compared to the integer order case. Due to this, a different kind of vector field analysis is required for characterizing the contraction behavior in the whole space. Mathematically, the above discussion can be expressed like this

$$ {}_0I_t^\alpha f(t) = \frac{1}{\Gamma(\alpha)} \int_0^t \frac{f(\tau)}{(t-\tau)^{(1-\alpha)}} d\tau, $$

One can write in form

$$ {}_0I_t^\alpha f(t) = \int_0^t f(\tau) dg_t(\tau) $$

$$ g_t(\tau) = \frac{1}{\Gamma(\alpha)} \left\{ t^\alpha - (t-\tau)^\alpha \right\}. $$

The scaling property of $g_t(\tau)$ is given as

$$ g_{t_1}(\tau_1) = g_{kt}(k\tau) = k^\alpha g_t(\tau). $$

when $t_1 = kt$ and $\tau_1 = k\tau$. Due to this scaling property, the closed loop homogeneous vector field $v_{\text{fractional}}$ of this fractional order system is modified as

$$ v_{\text{fractional}} = \frac{1}{\beta_1} x_1 \frac{\partial^\alpha}{\partial x_1^\alpha} + \cdots + \frac{1}{\beta_n} x_n \frac{\partial^\alpha}{\partial x_n^\alpha}. $$

where $\beta_n; n = 1, \ldots, n$ are defined as in (9.21). Moreover, the above vector field is linear with the Hurwitz (in fractional sense) characteristic polynomial $\lambda^n + k_n \lambda^{n-1} + \cdots + k_2\lambda + k_1$. Therefore, a global uniform contraction region is obtained after applying the proposed control to the fractional chain of integrators. □

9.6 Numerical Result

As an illustration of the results presented in this chapter, a commensurate fractional order linear model of the long Aluminum rod heated from one of its sides is considered in [14]. The input $u(t)$ to such a model is the thermal flux applied at one end of the rod and the output is the actual temperature at a prescribed section of the rod. Consider the continuous fractional order state-space model of the specified system, which is given by

$$ D^{0.5} x(t) = Ax(t) + bu(t), \tag{9.26} $$

where

$$ A = \begin{bmatrix} 0 & 1 & 0 \\ 0 & 0 & 1 \\ 0 & -0.0601251 & -0.42833 \end{bmatrix}, \quad b = \begin{bmatrix} 0 \\ 0 \\ 1 \end{bmatrix}. \tag{9.27} $$

The proposed controller for the above system (9.26) is represnted as $u = u_1 - A_0 x(t)$, where

$$A_0 = \begin{bmatrix} 0 & 0 & 0 \\ 0 & 0 & 0 \\ 0 & -0.0601251 & -0.42833 \end{bmatrix}. \tag{9.28}$$

After applying the control $u = u_1 - A_0 x(t)$, in (9.26), the closed loop system can be represnted as the fractional chain of integrtors

$$\begin{cases} D^{\alpha} x_1 &= x_2 \\ D^{\alpha} x_2 &= x_3 \\ D^{\alpha} x_3 &= u_1. \end{cases} \tag{9.29}$$

Now u_1 is given by (9.20)

$$u_1 = \left[\sum_{i=1}^{3} -k_i |x_i|^{\beta_i} sgn(x_i) - \sum_{j=1}^{3} \frac{(x_j) t^{\alpha-1}}{\Gamma(\alpha)} \right] \Omega, \tag{9.30}$$

where k_1, k_2 and k_3 are given by following polynomial

$$\lambda^3 + 0.78\lambda^2 + 0.4954\lambda + 0.7455 = 0. \tag{9.31}$$

Since all the coefficients of the above polynomial are positive, it is easy to check that $D(P) < 0$. Hence, the above polynomial is Routh Hurwitz in the fractional sense. $\beta_1 = 1/2, \beta_2 = 3/5, \beta_3 = 3/5$ and the initial conditions of state vector were chosen as $x_1 = \Gamma(0.5)$, $x_2 = -\Gamma(0.5)$, $x_3 = 1.5\Gamma(0.5)$. The numerical simulation for the states is depicted in Fig. 9.1.

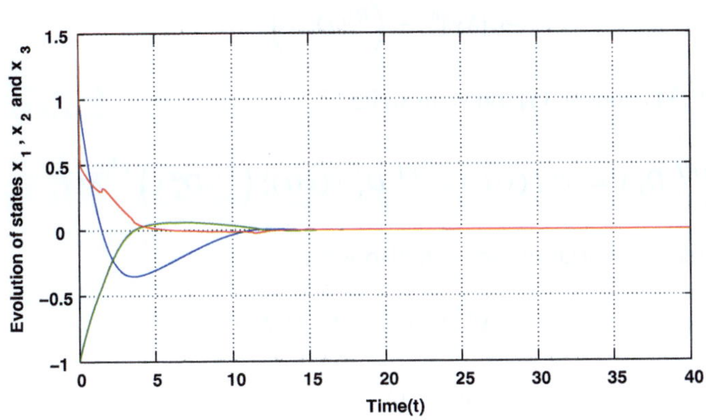

Fig. 9.1 Evolution of state (x_1, x_2 and x_3) w.r.t. time

9.7 Contraction Analysis by Fractional Order Infinitesimal Variation

Autonomous dynamical system (9.7) can be expressed in the following form

$$\dot{x} = {}^{RL}D_t^1 x = f(x(t)), \tag{9.32}$$

where f is a nonlinear vector field and $x(t)$ is an n-dimensional state vector. It is assumed that all the quantities are real and not necessarily smooth. Also, all the required Riemann-Liouville fractional derivatives exist and are continuous. In contraction analysis, the state x is fractionally modified to see the change in velocity vector $\dot{x}$, that is termed as the virtual displacement (fractional infinitesimal displacement at fixed time) $\delta^\alpha x$ and virtual velocity $\delta^\alpha \dot{x}$ respectively. Due to the first order variation of the state, virtual dynamics are introduced into the system, which can be represented by

$$\delta\dot{x} = \delta f(x(t)) = {}^{RL}D_x^1 f(x(t))\delta x. \tag{9.33}$$

where ${}^{RL}D_x^1 f(x(t)) = \frac{\partial f(x(t))}{\partial x}$. Similarly, due to the fractional infinitesimal displacement of the state, fractional virtual dynamics are introduced into the system, which can be represented by,

$$\begin{aligned} \delta^\alpha\, {}^{RL}D_t^1 x &= \delta^\alpha f(x(t)) = {}^{RL}D_x^\alpha f(x(t))[\delta x]^\alpha \\ [\delta x]^\alpha &= (\mathrm{sgn}(x))^m |\delta x|^\alpha. \end{aligned} \tag{9.34}$$

where $(m-1) < \alpha < m$, where m is an integer. For simplifying Eq. (9.34), we have to take the fractional variation of the variable x

$$\begin{aligned} \delta^\alpha x &= {}^{RL}D_x^\alpha x[\delta x]^\alpha \\ \Rightarrow [\delta x]^\alpha &= \left({}^{RL}D_x^\alpha x\right)^{-1} \delta^\alpha x. \end{aligned} \tag{9.35}$$

Using Eq. (9.34) and (9.35), one can write

$$\delta^\alpha\, {}^{RL}D_t^1 x = \delta^\alpha f(x(t)) = {}^{RL}D_x^\alpha f(x(t)) \left({}^{RL}D_x^\alpha x\right)^{-1} \delta^\alpha x. \tag{9.36}$$

By the property of variation one can also write

$$\delta^\alpha\, {}^{RL}D_t^1 x = {}^{RL}D_t^1 \delta^\alpha x. \tag{9.37}$$

Time derivative of squared distance $((\delta^\alpha x)^T \delta^\alpha x)$ between the two neighboring trajectories will be

$$^{RL}D_t^1((\delta^\alpha x)^T\delta^\alpha x) = 2(\delta^\alpha x)^{T\,RL}D_t^1(\delta^\alpha x). \tag{9.38}$$

Using Eqs. (9.36), (9.37) and (9.38), one can write

$$\begin{aligned} ^{RL}D_t^1((\delta^\alpha x)^T\delta^\alpha x) &= 2(\delta^\alpha x)^T \underbrace{^{RL}D_x^\alpha f(x(t))(^{RL}D_x^\alpha x)^{-1}}\delta^\alpha x \\ &\leq 2\lambda_m(x,t)(\delta^\alpha x)^T\delta^\alpha x \\ \Rightarrow \|\delta^\alpha x\| &\leq \|\delta^\alpha x_0\| e^{\int_0^t \lambda_m(x,t)dt}. \end{aligned} \tag{9.39}$$

where $\lambda_m(x,t)$ is the largest eigenvalues of the symmetric part of the

$$^{RL}D_x^\alpha f(x(t))(^{RL}D_x^\alpha x)^{-1}.$$

If $\lambda_m(x,t)$ is strictly uniformly negative, then any infinitesimal length $\|\delta^\alpha x\|$ converges exponentially to zero. Path integration in (9.39), ensures that all the solution trajectories of the system in (9.32) converge exponentially to a single trajectory, irrespective of initial conditions.

Definition 9.12 Given the system equations $\dot{x} = f(x,t)$, a region of the state space is called a contraction or semi-contraction region, if the matrix $^{RL}D_x^\alpha f(x(t))$ $\left(^{RL}D_x^\alpha x\right)^{-1}$ is uniformly negative definite or negative semi-definite in that region.

9.7.1 More Discussion About Contraction Region Using Coordinates Transformation

A locally state dependent nonsingular transformation Θ

$$\delta^\alpha z = \Theta\delta^\alpha x(t), \tag{9.40}$$

transforms the virtual dynamics (9.33) into $\delta^\alpha z$-coordinates as

$$\begin{aligned} ^{RL}D_t^1\delta^\alpha z &= {}^{RL}D_t^1\Theta\delta^\alpha x(t) + \Theta\,^{RL}D_t^1\delta^\alpha x(t) \\ &= {}^{RL}D_t^1\Theta\delta^\alpha x(t) \\ &\quad + \Theta\,^{RL}D_x^\alpha f(x(t))(^{RL}D_x^\alpha x)^{-1}\delta^\alpha x(t) \\ &= F'\delta^\alpha z, \end{aligned} \tag{9.41}$$

where F' is given by

$$\begin{aligned} F' &= \left(^{RL}D_t^1\Theta\delta^\alpha x(t) + \Theta\,^{RL}D_x^\alpha f(x(t))(^{RL}D_x^\alpha x)^{-1}\right)\Theta^{-1} \\ &= \left(\dot{\Theta} + \Theta\,^{RL}D_x^\alpha f(x(t))(^{RL}D_x^\alpha x)^{-1}\right)\Theta^{-1}. \end{aligned} \tag{9.42}$$

The rate of change of squared length is given by

$$\frac{d}{dt}(\delta^{\alpha}z^{T}\delta^{\alpha}z) = 2(\delta^{\alpha}z)^{T}F'\delta^{\alpha}z. \tag{9.43}$$

A more general definition of infinitesimal length can be given by

$$(\delta^{\alpha}x)^{T}(t)M(x(t))\delta^{\alpha}x(t), \tag{9.44}$$

where $M(x(t)) = \Theta^{T}\Theta$ is a symmetric, uniformly positive definite and continuously differentiable metric (formally, this defines a Riemannian manifold). Using the more general definition of infinitesimal length, one can calculate its rate of change as

$$\frac{d}{dt}((\delta^{\alpha}x(t))^{T}M(x(t))\delta^{\alpha}x(t)) = (\delta^{\alpha}x(t))^{T}\Xi\delta^{\alpha}x(t) \tag{9.45}$$

$$\begin{aligned}\Xi = &\left({}^{RL}D_{x}^{\alpha}f(x(t))({}^{RL}D_{x}^{\alpha}x)^{-1}\right)^{T}M \\ &+ M\,{}^{RL}D_{x}^{\alpha}f(x(t))\left({}^{RL}D_{x}^{\alpha}x\right)^{-1} + \dot{M}.\end{aligned} \tag{9.46}$$

Based on the above observation, definition of the contraction region is given as:

Definition 9.13 Given the n-dimensional system equations $\dot{x} = f(x(t))$, a region of the state space is called a contraction region with respect to a uniformly positive definite metric $M(x(t)) = \Theta^{T}\Theta$, if equivalently F' in Eq. (9.42) or $\Xi = \left({}^{RL}D_{x}^{\alpha}f(x(t))({}^{RL}D_{x}^{\alpha}x)^{-1}\right)^{T}M + M\,{}^{RL}D_{x}^{\alpha}f(x(t))({}^{RL}D_{x}^{\alpha}x)^{-1} + \dot{M}$ are uniformly negative definite in that region.

Remark 9.14 Even if the function is not differentiable at some finite number of points, one can check the stability by removing the discontinuity point.

Example Consider the autonomous differential equation

$$\dot{x} = -k\mathrm{sign}(x) = f(x), \tag{9.47}$$

where $k > 0$ and $\mathrm{sign}(x)$ defined as

$$\mathrm{sign}(x) = \begin{cases} +1 & \text{if } x > 0 \\ (-1\ \ 1) & \text{if } x = 0 \\ -1 & \text{if } x < 0 \end{cases} \tag{9.48}$$

For checking the stability of system (9.47), calculating the value of

$$\left({}^{RL}D_{x}^{\alpha}f(x(t))\right)\left({}^{RL}D_{x}^{\alpha}x\right)^{-1}$$

$$\begin{aligned}({}^{RL}D_x^{\alpha} - k\text{sign(x)})({}^{RL}D_x^{\alpha}x)^{-1} &= -\frac{k\text{sign}(x)x^{-\alpha}(\Gamma(2-\alpha))}{(\Gamma(1-\alpha))x^{1-\alpha}} \\ &= -\frac{k\frac{x}{|x|}(\Gamma(2-\alpha))}{(\Gamma(1-\alpha))x} \\ &= -\frac{k(\Gamma(2-\alpha))}{(\Gamma(1-\alpha))\,|x|} < 0. \end{aligned} \tag{9.49}$$

Hence any two trajectories of the above system is contracting.

Example Fractional order controller design for an integer order system Consider the following simplest integer order system

$$\dot{x} = u. \tag{9.50}$$

We have to design u based on convergence condition of contraction theory such as

$$({}^{RL}D_x^{\alpha}u)({}^{RL}D_x^{\alpha}x)^{-1},$$

is uniformly negative definite in the whole state space. There are infinitely many possible values of u, which satisfies the above convergence condition. One of the simplest possible value is $u = {}^{RL}D_x^{-\alpha}(-kx^{1-\alpha})$ where $k > 0$. Substituting the value of u in the convergence condition, one can get

$$\begin{aligned}({}^{RL}D_x^{\alpha}u)({}^{RL}D_x^{\alpha}x)^{-1} &= ({}^{RL}D_x^{\alpha}\,{}^{RL}D_x^{-\alpha}(-kx^{1-\alpha}))({}^{RL}D_x^{\alpha}x)^{-1} \\ &= \frac{(-kx^{1-\alpha})}{\frac{x^{1-\alpha}}{\Gamma(2-\alpha)}} \\ &= -k\Gamma(2-\alpha) < 0. \end{aligned} \tag{9.51}$$

Hence the proposed controller $u = {}^{RL}D_x^{-\alpha}(-kx^{1-\alpha})$ is able to stabilize the system (9.50).

9.8 Contraction Analysis of Fractional Order Systems Using the Proposed Strategy

Consider a fractional order dynamical system

$${}^{RL}D_t^{\alpha}x(t) = f(x(t)). \tag{9.52}$$

For applying contraction theory, taking ${}^{RL}D_t^{1-\alpha}$ in both sides of the Eq. (9.52), one can write,

$$^{RL}D_t^1 x(t) = \dot{x}(t) = {^{RL}D_t^{1-\alpha}} f(x(t)). \tag{9.53}$$

Definition 9.15 Given the fractional order systems $^{RL}D_t^\alpha x(t) = f(x, t)$, a region of the state space is called a contraction or semi-contraction region, if the matrix $^{RL}D_x^{\alpha}\, {^{RL}D_t^{1-\alpha}} f(x(t)))({^{RL}D_x^\alpha} x)^{-1}$ is uniformly negative definite or negative semi-definite in that region.

Example Consider the autonomous differential equation

$$^{RL}D_t^\alpha x = -k\text{sign}(x) = f(x) \tag{9.54}$$

where $k > 0$ and $\text{sign}(x)$ defined as

$$\text{sign}(x) = \begin{cases} +1 & \text{if } x > 0 \\ (-1 \;\; 1) & \text{if } x = 0 \\ -1 & \text{if } x < 0 \end{cases}$$

For checking the stability of system (9.54), calculating the value of

$$({^{RL}D_x^{\alpha}}\, {^{RL}D_t^{1-\alpha}} f(x(t)))({^{RL}D_x^\alpha} x)^{-1}$$

$$\begin{aligned} ({^{RL}D_x^{\alpha}}\, {^{RL}D_t^{1-\alpha}} k\text{sign(x)})({^{RL}D_x^\alpha} x)^{-1} &= -\frac{k\text{sign}(x)t^{\alpha-1}x^{-\alpha}(\Gamma(2-\alpha))}{(\Gamma(1-\alpha))(\Gamma(\alpha))x^{1-\alpha}} \\ &= -\frac{k\frac{x}{|x|}t^{\alpha-1}(\Gamma(2-\alpha))}{(\Gamma(1-\alpha))(\Gamma(\alpha))x} \\ &= -\frac{k(\Gamma(2-\alpha))t^{\alpha-1}}{(\Gamma(1-\alpha))(\Gamma(\alpha))\,|x|} < 0 \end{aligned} \tag{9.55}$$

Hence any two trajectories of the above fractional order system is contracting.

9.9 Summary

The exponential stability of a fractional order system with the proposed controller is proved using classical contraction analysis. It can be seen that when $\alpha = 1$, the controller (9.20) takes the following form:

$$\begin{aligned} u|_{\alpha=1} &= -(k_1 sgn(x_1)|x_1|^{\beta_1} + x_1) - (k_2 sgn(x_2)|x_2|^{\beta_2} + x_2) \\ &\quad - \cdots - (k_n sgn(x_n|x_n|^{\beta_n} + x_n). \end{aligned}$$

This controller is very similar to the one proposed by [12] which is discussed in Sect. 9.4. The only difference is the addition of the proportional term. This proportional term will accelerate the convergence of the system states.

Also, in this chapter contraction theory concept has been revisited, replacing the first order variation of system states with the fractional order variation. Classical contraction theory has been able to analyze the convergence behavior of nonlinear systems in state space only when system is continuously differentiable in nature. The main advantage of the proposed approach is that it is also useful for analyzing the stability of non differentiable systems. Also, this approach is useful for analyzing stability of fractional order systems and designing the fractional order controller for dynamical system described by integer order differential equation. Numerical examples have been presented to illustrate the proposed method.

References

1. Lohmiller, W., Slotine, J.J.E.: On contraction analysis for non-linear systems. Automatica **34**(6), 683–696 (1998)
2. Aylward, E.M., Parrilo, P.A., Slotine, J.J.E.: Stability and robustness analysis of nonlinear systems via contraction metrics and sos programming. Automatica **44**(8), 2163–2170 (2008)
3. Jouffroy, J., Fossen, I.: A tutorial on incremental stability analysis using contraction theory. Model. Identif. Control **31**(3), 93–106 (2010)
4. Angeli, D.: A lyapunov approach to incremental stability properties. IEEE Trans. Autom. Control **47**(3), 410–421 (2002)
5. Kamal, S., Bandyopadhyay, B., Spurgeon, S.: Stabilization of a fractional-order chain of integrators: a contraction-based approach. IMA J. Math. Control Inf., dnt042 (2013)
6. Tarasov, V.E.: Fractional variations for dynamical systems: hamilton and lagrange approaches. J. Phys. A Math. Gen. **39**(26), 8409 (2006)
7. Tarasov, V.E.: Fractional derivative as fractional power of derivative. Int. J. Math. **18**(03), 281–299 (2007)
8. Chen, Y.Q., Petras, I., Xue, D.: Fractional order control—a tutorial. In: American Control Conference, 2009. ACC'09., pp. 1397–1411. IEEE (2009)
9. Ahmed, E., El-Sayed, A., El-Saka, H.A.: On some Routh-Hurwitz conditions for fractional order differential equations and their applications in Lorenz, rössler, Chua and Chen systems. Phys. Lett. A **358**(1), 1–4 (2006)
10. Mishina, A., Proskuryakov, I.: Higher Algebra. Fizmatgiz, Moscow (1965)
11. Podlubny, I.: Fractional Differential Equations: An Introduction to Fractional Derivatives, Fractional Differential Equations, to Methods of Their Solution and Some of Their Applications, vol. 198. Access Online via Elsevier (1998)
12. Bhat, S.P., Bernstein, D.S.: Geometric homogeneity with applications to finite-time stability. Math. Control Signals Syst. **17**(2), 101–127 (2005)
13. Djamah, T., Mansouri, R., Bettayeb, M., Djennoune, S.: State space realization of fractional order systems. In: Intelligent Systems & Automation: 2nd Mediterranean Conference on Intelligent Systems and Automation (CISA09), vol. 1107, pp. 37–42. AIP Publishing (2009)
14. Sabatier, J., Agrawal, O.P., Machado, J.T.: Advances in fractional calculus: theoretical developments and applications in physics and engineering. Springer, Dordrecht (2007)

Index

B. Bandyopadhyay and S. Kamal, *Stabilization and Control of Fractional Order Systems: A Sliding Mode Approach*, Lecture Notes in Electrical Engineering 317,
DOI 10.1007/978-3-319-08621-7

MIX
Papier aus verantwortungsvollen Quellen
Paper from responsible sources
FSC® C105338

If you have any concerns about our products,
you can contact us on
ProductSafety@springernature.com

In case Publisher is established outside the EU,
the EU authorized representative is:
Springer Nature Customer Service Center GmbH
Europaplatz 3, 69115 Heidelberg, Germany

Printed by Libri Plureos GmbH
in Hamburg, Germany